Extreme Space Weather

Extreme Space Weather

RYUHO KATAOKA
National Institute of Polar Research, JAPAN
The Graduate University for Advanced Studies, SOKENDAI

ELSEVIER

Elsevier
Radarweg 29, PO Box 211, 1000 AE Amsterdam, Netherlands
The Boulevard, Langford Lane, Kidlington, Oxford OX5 1GB, United Kingdom
50 Hampshire Street, 5th Floor, Cambridge, MA 02139, United States

ISBN: 978-0-12-822537-0

For Information on all Elsevier publications visit our website at
https://www.elsevier.com/books-and-journals

Publisher: Candice Janco
Acquisitions Editor: Peter Llewellyn
Editorial Project Manager: Jose Paolo Valeroso
Production Project Manager: Sruthi Satheesh
Cover Designer: Mark Rogers

Typeset by Aptara, New Delhi, India

Contents

Preface *vii*
Abbreviations and acronyms *ix*

1. Introduction to space weather **1**

1.0 Introduction 1
1.1 Sunspots and solar wind 1
1.2 Geomagnetic field 12
1.3 Atmosphere 19

2. Disturbed space weather **31**

2.1 Solar flares 31
2.2 Coronal mass ejections 35
2.3 Solar energetic particles 39
2.4 Geomagnetic disturbances (GMD) 45
2.5 Radiation belts 56

3. Technological vulnerability and statistics **65**

3.1 Effects of solar flares 65
3.2 Effects of geomagnetic disturbances (GMD) 67
3.3 Effects of energetic particles 76
3.4 Statistics of extreme events 80
3.5 Outstanding space weather events 87
3.5.1 September 1–2, 1859: The Carrington event 87
3.5.2 May 14, 1921: The New York Railroad event 88
3.5.3 February 23, 1956: The greatest GLE event 88
3.5.4 February 11, 1958: The Tokyo aurora event 88
3.5.5 August 4, 1972: The high-speed and worst SEP event 89
3.5.6 March 13–14, 1989: The Hydro-Quebec blackout event 89
3.5.7 September–October, 1989: The MIR event 89
3.5.8 March 24, 1991: The super SC event 90
3.5.9 July 15, 2000: The Bastille Day event 90
3.5.10 October–November, 2003: The Halloween event 90
3.5.11 January 20, 2005: The second largest GLE event 91
3.5.12 July 23, 2012: The STEREO event 92
3.5.13 September 10, 2017: The Mars event 92

4. Forecasting space hazards **93**

4.1 Forecasting solar flares 93
4.2 Arrival of coronal mass ejections 96
4.3 Forecasting substorms 99
4.4 Forecasting magnetic storms 102
4.5 Forecasting radiation belts 106
4.6 Forecasting ground level enhancements 111
4.7 Metrics for evaluating different forecast models 113

5. Toward the Moon and Mars **115**

5.1 Predicting galactic cosmic rays (GCR) 115
5.2 Predicting solar energetic particles (SEP) 120
5.3 The Moon 125
5.4 Mars 131

References *139*
Index *169*

Preface

Where are we in the Universe? This is a basic question to begin learning about the space weather. We are living with a star, known as the Sun. This is a simple answer, including many fruitful insights. Is our planet Earth safe enough to foster life and society, regardless of the close location to the Sun? Some other specific questions then naturally arise. What kind of shields is the Earth equipped with? What kind of spears does the Sun have? What are vulnerable infrastructures? Although the distance between the Sun and the Earth remains almost constant, temporal variations of the Sun and Earth themselves increase the complexity of the relationships. To understand the space weather, the whole spectrum of space physics must be studied; however, fluid mechanics and electromagnetism are the essential topics in physics that should be known. Galactic cosmic rays, solar energetic particles, and auroras share the same principles of physics, and the borders among these fields of research are not clear. This field of study is now called heliophysics.

What do you answer if you are asked to explain the mechanism of auroras? Specifically, can you explain how to forecast auroral activity 1 hour, 1day, or 1 month in advance based on the available observational data and simulations? The answers to these questions cover a large part of essential aspects to understand the space weather, as auroras are caused by the interactions among three elements: The solar wind, geomagnetic field, and atmosphere. This small book is a trial for the author, an auroral scientist, to concisely introduce the complex situation of the cosmic shore.

Auroras are usually high-latitude phenomena, but they can appear anywhere in the world. When a giant magnetic storm occurs, it may send electric currents to mid- and low-latitude areas where auroras do not appear otherwise. These auroras must be in reddish hues, and the "red sign" has been historically observed even in mid-latitude Japan. Such extreme magnetic storms influence the sustainability of modern high-tech society, which will be discussed later in this book.

It has been always challenging to predict and mitigate every aspect of natural hazards, such as volcanic eruptions, earthquakes, and typhoons. Prediction of severe space weather, such as solar eruptions and magnetic storms, is also still challenging. However, accumulating the fundamental knowledge on extreme events is useful to prepare against future space hazards. This book is based on the knowledge collected by the author during the latter half of solar cycle 23 (1996–2009), the complete solar cycle 24 (2009–2019), and the beginning of solar cycle 25 (2020~). The upcoming new solar cycles will also provide novel interesting information about the space weather, which will be necessary for future expeditions to the Moon and Mars.

In this age of internet, everyone can find data sources including great footages of solar eruptions to auroral breakup as well as highly specialized knowledge of space physics. In this circumstance, this small book aims to quickly draw a big picture for learning heliophysics. Extreme space weather is the keyword to achieve that purpose. Such an approach is a natural consequence of the "extreme space weather" workshops held in Japan in 2013 to 2015, which published unique scientific papers that became the foothold of this book.

The main audience of this book is supposed to be Science and Engineering graduate students. The theoretical backbone of this book is based on the lectures given for graduate students, who are already familiar with basic fluid dynamics and electromagnetism. However, this book would also be useful for broad space weather audience, including decision makers, rapidly growing space industry leaders, power grid specialists, and mission analysts for long-term manned space missions, to prepare for space hazards.

The author would like to thank Tatsuhiko Sato, Kiyoka Murase, Yoshizumi Miyoshi, Daikou Shiota, Keisuke Hosokawa, Yuki Kubo, Masaki Nishino, Hiromu Nakagawa, Hiroko Miyahara, Emilia Kilpua, Louis J. Lanzerotti, and Emile Touber for their kind help to complete this book. The cover image is provided from auroral photographer Shiori Uchino.

Ryuho Kataoka
Sabbatical visit at Okinawa Institute of Science and Technology,
October 15, 2021.

Abbreviations and acronyms

CIR	Corotating interaction region
CME	Coronal mass ejection
CPCP	Cross polar cap potential
EVA	Extravehicular activity
FAC	Field-aligned current
GCR	Galactic cosmic rays
GEO	Geosynchronous orbit
GIC	Geomagnetically induced currents
GLE	Ground level enhancement (of neutron count rates)
GMC	Geosynchronous magnetopause crossing
GMD	Geomagnetic disturbance
GNSS	Global Navigation Satellite System
HCS	Heliospheric current sheet
IMF	Interplanetary magnetic field
ISS	International Space Station
LEO	Low-earth orbit
MHD	Magnetohydrodynamics
NICT	National Institute of Information and Communications Technology
NASA	National Aeronautics and Space Administration
NOAA	National Oceanic and Atmospheric Administration
PCA	Polar cap absorption
REP	Relativistic electron precipitation
SAPS	Sub-auroral polarization stream
SBZ	Southward BZ (magnetic field of the solar wind)
SAA	South Atlantic Anomaly
SC	Sudden commencement
SEP	Solar energetic particles
SEE	Single event effect
SEU	Single event upset
SSC	Storm sudden commencement
SWPC	Space weather prediction center (at NOAA)
TEC	Total electron content
ULF	Ultra-low frequency

Physical constants and useful units

Light speed	c	2.99792458×10^{8} m s^{-1}
Plank constant	$\hbar$	$6.62607015 \times 10^{34}$ J s
Elementary charge	e	$1.602176634 \times 10^{-19}$ C
Boltzmann constant	k_B	1.380649×10^{23} J K^{-1}
Proton mass	m_p	$1.672621898 \times 10^{27}$ kg
Electron mass	m_e	$9.1093837015 \times 10^{31}$ kg
Permeability in vacuum	μ_0	$4\pi \times 10^{-7}$ N A^{-2}
Permittivity of vacuum	ε_0	$8.8541878128 \times 10^{-12}$ F m^{-1}
Earth radius	R_E	6,378 km
Solar radius	R_S	6.96×10^{8} m ~ 109 R_E
Astronomical unit	AU	1.496×10^{11} m ~ 215 R_S
Unit of energy	erg	10^{-7} J
Unit of wavelength	Å	0.1 nm
Unit of magnetic field	G	10^{5} nT
Unit of absorbed dose	Gy	100 rad
Unit of effective dose	Sv	100 rem

CHAPTER 1

Introduction to space weather

1.0 Introduction

It may be helpful to overview the surrounding world via the auroras first, before going into the details of other space weather phenomena. The near-Earth space environment causing auroras is called *geospace* (Fig. 1.1). The Sun always emits electromagnetic waves, such as X-ray, ultraviolet ray, and visible light, and also emits ionized charged particles called *plasma*. The supersonic *solar wind* plasma released from the Sun clash into the protective magnetic barrier of the Earth, which sparks a massive amount of electricity. The produced electric current travels toward the polar regions along the lines of magnetic force, forming a ring-shaped display, called the *auroral oval*. Tracing the solar wind particles from the Sun to the Earth fails to predict the occurrence of auroral oval because the solar wind particles do not hit the atmosphere of the Earth. Instead, global magnetohydrodynamic (MHD) simulations successfully identify the electric current flow in the geospace to predict the basic auroral oval distribution.

1.1 Sunspots and solar wind

The Sun is a magnetized astronomical body. During the total solar eclipse, these magnetic field lines can be observed as a beautiful coronal emission pattern through the naked eyes. It is the glowing plasma trapped in the magnetic field. A localized strong magnetic field spot sometimes appears and disappears on the solar surface, which is called *sunspot*. The sunspot is dark because it is cooler (~4000 K) than the surrounding (~6000 K), due to a 0.2 to 0.3 Tesla strong magnetic field against the quiet Sun magnetic field strength of <0.1 T. The sunspots live for days to weeks. The sunspot variation recorded by Galileo Galilei ~400 years ago initiated our approach to understand the space weather and space climate.

The solar luminosity is almost constant at 3.84×10^{33} erg/s, which shows only 0.1% variation, because the Sun is powered by the fusion process in the core ($<0.27\ R_{Sun}$). As shown in Fig. 1.2, the heat is conducted by radiation in the radiative zone ($<0.7\ R_{Sun}$), and convection in

Extreme Space Weather
DOI: https://doi.org/10.1016/B978-0-12-822537-0.00003-X

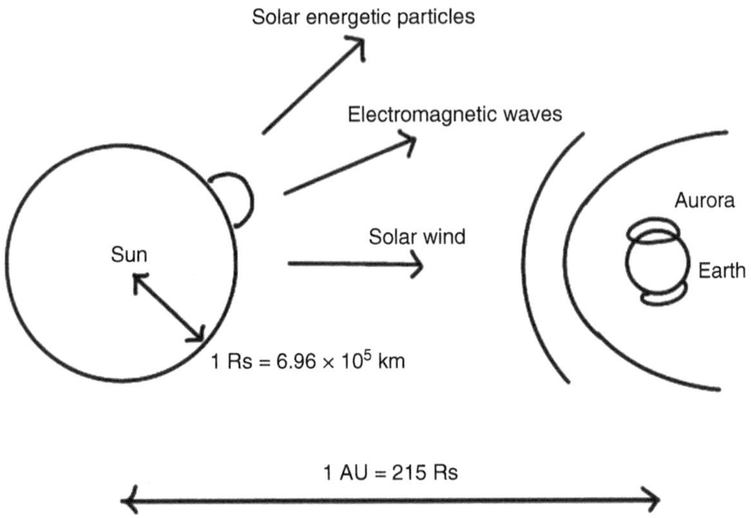

Fig. 1.1 ***Schematic illustration of the solar–terrestrial environment.***

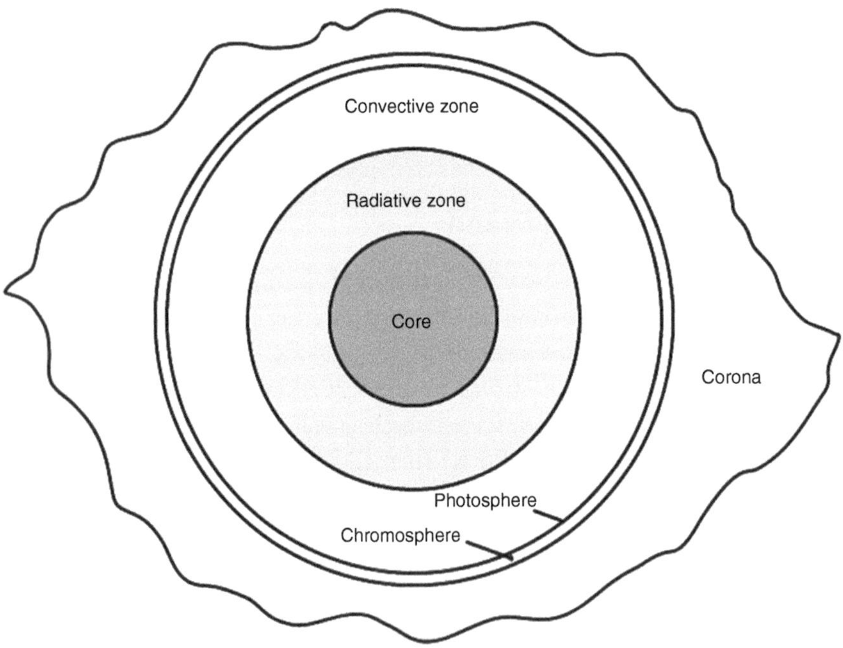

Fig. 1.2 ***The inner structure and the atmosphere of the Sun.***

the convective zone (0.7–1.0 R_{Sun}). The *photosphere*, also called the solar surface, is 100 km thick. The bubbling pattern in the photosphere is called *granules*. The red atmosphere is the *chromosphere*, a few thousand km thick layer located just above the photosphere. *Solar corona* is the extended atmosphere located approximately >2500 km above the photosphere. From the chromosphere to corona, the density drops from 10^{11} /cc to 10^9 /cc, while temperature increases from 10^4 K to 10^6 K. Since Covington (1947) discovered the connection between the sunspots and centimetric radio emissions, the 10.7 cm solar radio flux (F10.7) became a widely used index. In fact, the 10.7 cm wavelength (2.8 GHz frequency) is appropriate for monitoring solar activity level because it is sensitive to conditions in the upper chromosphere and at the base of the corona (Tapping, 2013).

The Sun rotates unevenly. The rotation period of the Sun is ranging from 34 days at poles and 25 days at equator. The meridional circulation speed at the photosphere is ~10 m/s. Both rotation and circulation act as the dynamo to produce the strong solar magnetic field (Hotta and Kusano, 2021). The cyclicity helps understanding the laws of nature. The most famous cyclicity on the solar surface is the 11-year sunspot cycle called *Schwabe cycle*, where the number of sunspots increase and decrease after approximately every 11 years. The negative and positive peaks of the sunspot time series are called *solar minimum* and *solar maximum*, respectively (Fig. 1.3). The space environment is relatively steady when few sunspots appear, and relatively active when many sunspots emerge. The magnetic moment polarity of the Sun flips every solar maximum, so that it takes almost 22 years to return to the identical situation. This 22-year cycle is called *Hale cycle*.

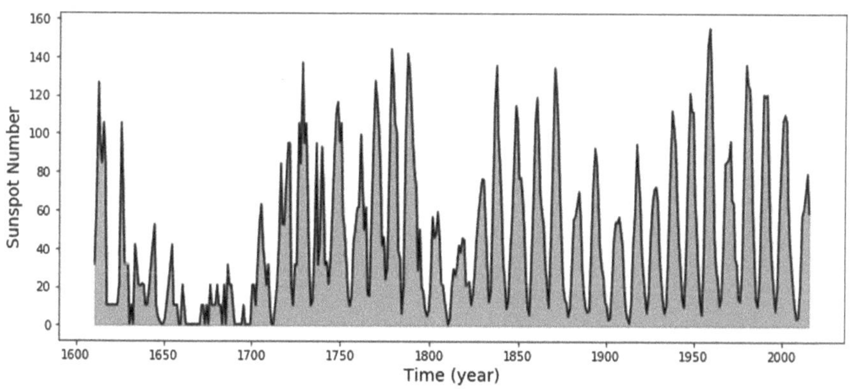

Fig. 1.3 ***Sunspot number as reconstructed by Svalgaard and Schatten (2016).***

The physical mechanism to repeat the magnetic cycle is unsolved. It may be understood by the basic process to release the accumulated magnetic energy from inside the Sun to outside. Turbulent convection in the convective zone essentially involves the generation of localized strong magnetic field everywhere (Hotta et al., 2015). The strong magnetic field must emerge beyond the photosphere at some points by the magnetic buoyancy to store the magnetic energy in the solar corona. The most efficient ways to release the coronal magnetic energy are two-fold, as described in Chapter 2, by converting a part of the magnetic energy into heat (solar flares) and ejecting the magnetized plasma itself to the solar wind (coronal mass ejections).

The first example of extreme space climate is the Maunder Minimum, when the sunspots mostly disappeared for 70 years from 1645 (Fig. 1.3). The number of auroral witness records over the world also decreased during this time (Usoskin et al., 2015). Such an extremely weakened long-term solar activity is called *grand minimum*. The 11-year cyclicity itself has been identified using other proxies, such as cosmogenic ^{10}Be in ice cores (Beer et al., 1998) and ^{14}C content in tree rings (Miyahara et al., 2004), even during the Maunder Minimum. The possible causes and consequences of the Maunder Minimum are still uncertain and has been actively discussed (Miyahara et al., 2021). Brehm et al. (2021) summarized the 11-year solar cycles over the last millennium.

A part of the coronal magnetic field is open, extending outward and streaming away with the *interplanetary magnetic field* which is embedded in the *solar wind*. The solar wind is continuous but not uniformly flowing out from the Sun. The Ulysses spacecraft (1990–2009) revealed two different types of the solar wind, fast and slow solar wind, through the latitudinal scan in the solar system (Fig. 1.4). The high-speed solar wind flowing out from *coronal hole* is the important phenomenon to prepare against space weather hazards in *radiation belts* of the Earth, although sunspots are the start-point for many other serious space weather situations as shown later in Chapter 2.

The fast and slow solar wind streams interact in the interplanetary space to form *corotating interaction region* (CIR). The discontinuous plasma boundary separating the fast and slow solar wind is called *stream interface*. In a steady state, the compressed structure corotates with the solar rotation (Fig. 1.5), and repeatedly hits the Earth every 27 days to cause recurrent *magnetic storms*, through its strong compressed magnetic field. CIRs are also called stream interaction regions (SIRs), if we do not distinguish the structure is corotating or not.

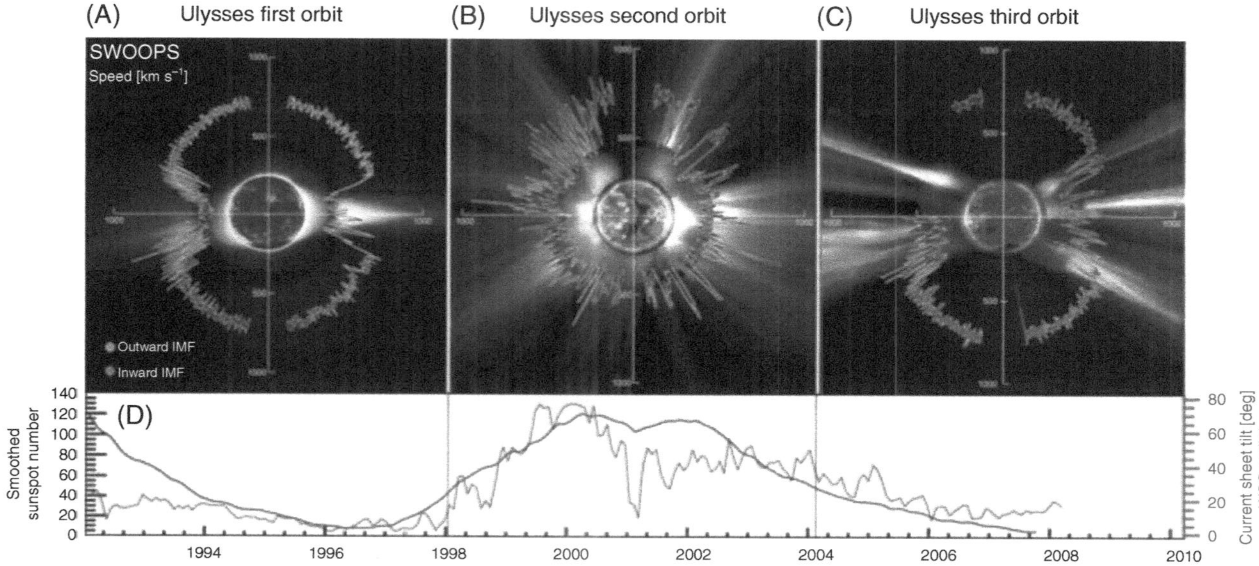

Fig. 1.4 ***The latitudinal variation of solar wind speed measured by the Ulysses spacecraft, during the solar minimum and solar maximum.*** The blue and red colors indicate the polarity of the magnetic field, which was flipped in the first and third orbit (McComas et al., 2003, 2008).

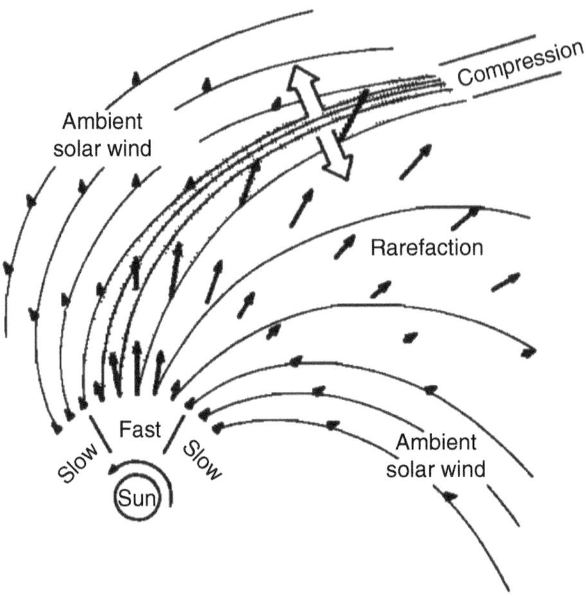

Fig. 1.5 ***Schematic illustration of the Parker spirals from the fast and slow winds, and the corotating interaction region (CIR) where the secondary nonradial motion (large open arrows) is caused by the pressure gradients in the stream interaction (Pizzo, 1978).***

MHD simulation explains the macroscopic physics of the corona and the solar wind because these gases are plasma, that is, hot enough to be ionized and magnetized.

The motion of plasma causes electric field and electric currents, and the electromagnetic force changes the plasma flow. We can therefore introduce momentum and mass conservation equations as follows:

$$\rho \frac{d\mathbf{v}}{dt} = \mathbf{j} \times \mathbf{B} - \nabla p + \rho \mathbf{G} \tag{1.1}$$

$$\frac{\partial \rho}{\partial t} = -\nabla \cdot (\rho \mathbf{v}) \tag{1.2}$$

where $\mathbf{j}$ is the electric current and $\mathbf{B}$ is the magnetic field, while ρ, $\mathbf{v}$, and p indicate mass density, velocity, and gas pressure, respectively. The gravity $\rho\mathbf{G}$ was added to Eq. (1.1) as an example of the external force. The Maxwell equations can be practically simplified as:

$$\nabla \times \mathbf{B} = \mu_0 \mathbf{j} + \frac{1}{c^2}\frac{\partial \mathbf{E}}{\partial t} \tag{1.3}$$

$$\nabla \times \mathbf{E} = -\frac{\partial \mathbf{B}}{\partial t} \tag{1.4}$$

$$\nabla \cdot \mathbf{B} = 0 \tag{1.5}$$

$$\mathbf{j} = \sigma\left(\mathbf{E} + \mathbf{v} \times \mathbf{B}\right) \tag{1.6}$$

where $\mathbf{E}$ is electric field, and σ is the electric conductivity.

If the plasma is isotropic and adiabatic, $p \propto \rho^{\gamma}$, where γ is the specific heat ratio. From this equation set, Alfvén (1942) first suggested that the combined electromagnetic and hydrodynamic waves exist as the fundamental mode in the conducting fluid, which is now known as *Alfvén waves*. Fig. 1.6 shows an example of the Alfvén waves identified in the solar wind.

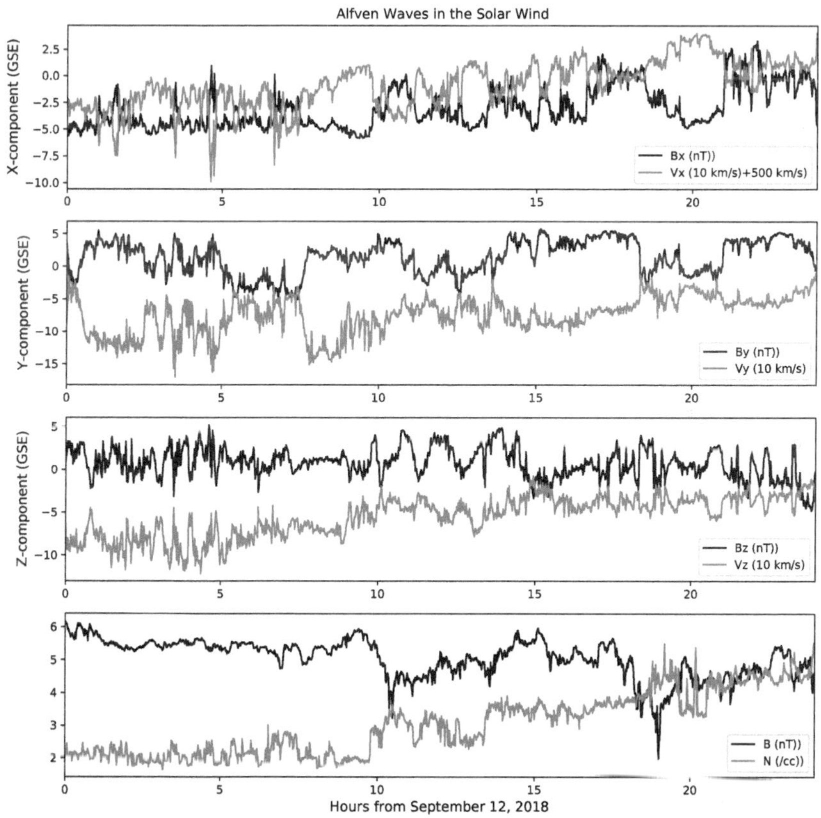

Fig. 1.6 ***The Alfvén waves in the solar wind, as identified by a good correlation between the vector fluctuations in magnetic field and velocity.*** The pioneering work was given by Belcher and Davis (1971).

The Alfvén waves propagating from the photosphere, and dissipating their energy in the corona, can explain both the hot corona and supersonic solar wind, as shown by recent MHD simulations (Suzuki and Inutsuka, 2006; Suzuki et al., 2013; Matsumoto and Suzuki, 2014; Shoda et al., 2018).

From Eq. (1.2), assuming the radial expansion of the solar wind plasma at a constant speed u in a spherical symmetry, the *dynamic pressure* of the solar wind applied to the geomagnetic field at 1 AU position can be expressed as follows:

$$p_{\mathrm{d}} = 1.3\left(\frac{r}{1\,\mathrm{AU}}\right)^{-2}\left(\frac{n_0}{5/\mathrm{cc}}\right)\left(\frac{u}{400\,\mathrm{km/s}}\right)^{2}\ \mathrm{nPa}, \tag{1.7}$$

where n_0, the solar wind proton number density, was assumed to be 5/cc at 1 AU.

The most popular assumption as derived from the ideal MHD equations is so-called *frozen-in* condition where the conductivity $\sigma \rightarrow \infty$ and $\mathbf{E} = -\mathbf{v} \times \mathbf{B}$ in the Ohm's law of Eq. (1.6), so that the plasma inside a magnetic flux tube moves together. In the reference frame of the solar system, the Sun and corona rotates, and the extended magnetic field by the solar wind shapes a spiral structure, which is known as *Parker spiral* (Parker, 1958). Assuming the frozen-in condition and the constant solar wind speed u, the radial expansion of the solar wind gives the radial component of magnetic field as:

$$B_r = B_0\left(\frac{r}{r_0}\right)^{-2}, \tag{1.8}$$

where B_0 is the radial magnetic field at the radial distance $r = r_0$. An average rotation period of the Sun is 25.4 days near the solar equator; thus, the angular velocity ω is 2.7×10^{-6} rad/s. The azimuthal component is therefore:

$$B_\varphi = -\frac{\omega r}{u} B_r \tag{1.9}$$

The equation to draw a field line is:

$$r - r_0 = -\frac{u}{\omega}(\varphi - \varphi_0), \tag{1.10}$$

where, the location of the source of a field-line is at longitude $\varphi = \varphi_0$ at radial distance $r = r_0$.

The radial components of the *interplanetary magnetic field* (IMF) extending from northern hemisphere and southern hemisphere of the Sun have different polarities. The north–south solar wind regions with different

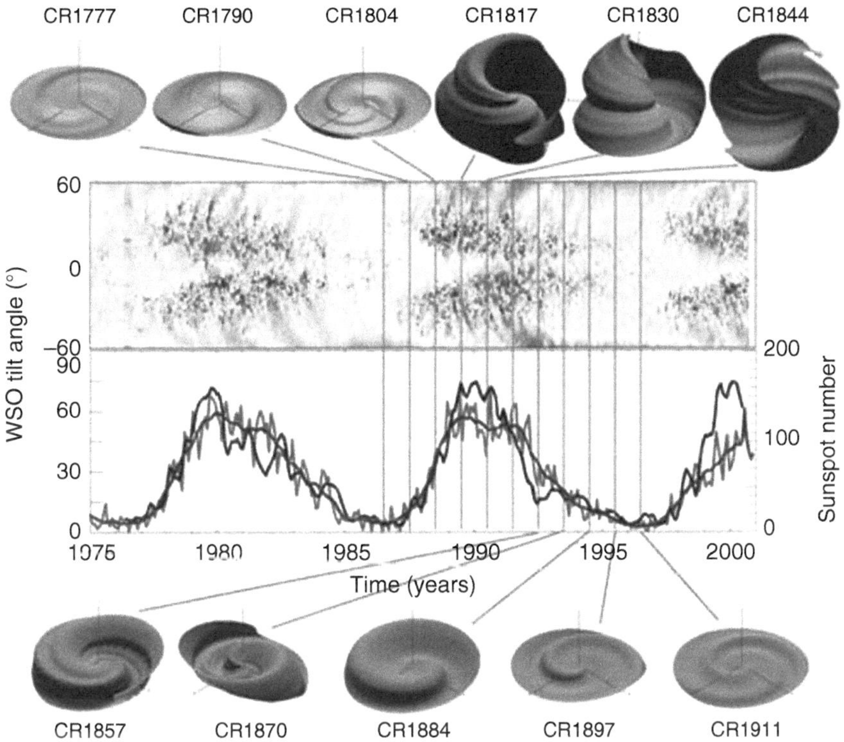

Fig. 1.7 ***Evolution of the heliospheric current sheet (HCS) during solar cycle 22.*** The "butterfly diagram" of sunspots location, sunspot number, and the tilt angle of the magnetic neutral line are also shown in the middle panels (Riley et al., 2002).

polarity are separated by the *heliospheric current sheet* (HCS). The shape of HCS evolves as shown in Fig. 1.7. The HCS extends toward the *heliopause*, where the solar wind plasma meets to the interstellar medium, across the *termination shock*, where the solar wind feels the obstacle. Adjacent to and beyond the heliopause, there is a region of heated and compressed neutral interstellar hydrogen, called *hydrogen wall* (Zank, 1999). The schematic illustration of the heliosphere is shown in Fig. 1.8. The actual heliospheric shape can be croissant-like (Opher et al., 2020).

The heliopause distance can be roughly estimated from the density (~0.1 /cc) and relative speed (~26 km/s) of the interstellar medium. Eq. (1.7) shows that the dynamic pressure of the solar wind at 100 AU is in order of 0.1 pPa, which is comparable to that from the interstellar medium. In fact, Voyager 1 passed across the heliopause on August 25, 2012 at ~122 AU (Burlaga et al., 2013; Krimigis et al., 2013; Stone et al., 2013). Voyager 2 crossed

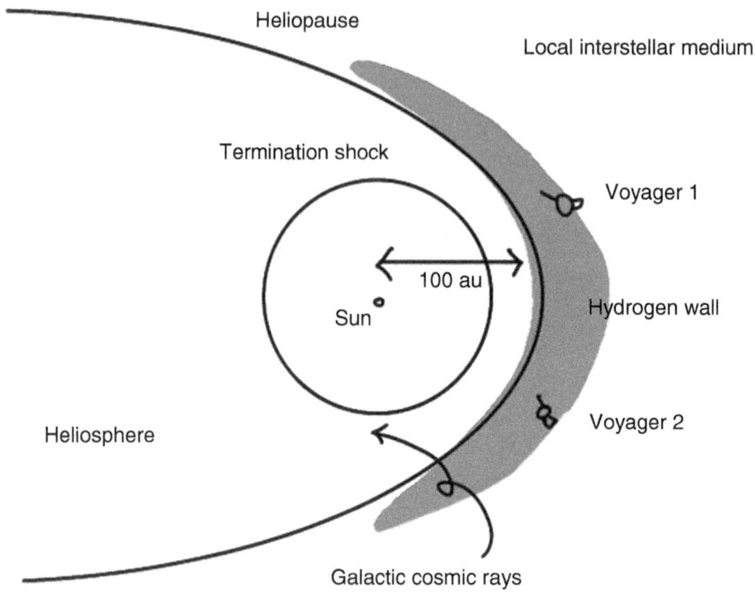

Fig. 1.8 ***Schematic illustration of the heliosphere.*** All of the planets are orbiting inside of the termination shock.

the heliopause in November 2018 at ~119 AU (Dialynas et al., 2019), which provides information on the interstellar neutral hydrogen density (~0.12 /cc) and a magnetic field upstream of the heliopause (~0.5 nT).

The global evolution of HCS affects the background radiation environment of the Earth, through high-energy charged particles called *galactic cosmic rays* (GCR). The GCR come from outside the heliosphere and intrude it against the magnetic field structure of the whole heliosphere. The time evolution of the magnetic barrier largely depends on the sunspot cycle (Fig. 1.7). The 11-year solar modulation of GCR will be introduced in more detail in Chapter 5. The energy spectra of GCR follow a power-law distribution from GeV to PeV, with a single spectral index of ~2.7 (Fig. 1.9, Particle Data Group, 2020). In detail, however, Calorimetric Electron Telescope (CALET) onboard the International Space Station (ISS) observed breaks in the spectral index of 2.87 and 2.56 at ~100 and ~500 GeV, respectively (Adriani et al., 2019).

Supernovae occur approximate once every 100 years in our galaxy to supply GCR with such a wide energy range. The *diffusive shock-acceleration* (Bell, 1978; Blandford and Ostriker, 1978) of charged particles at the blast wave of supernova remnant is a standard theory to explain the basic

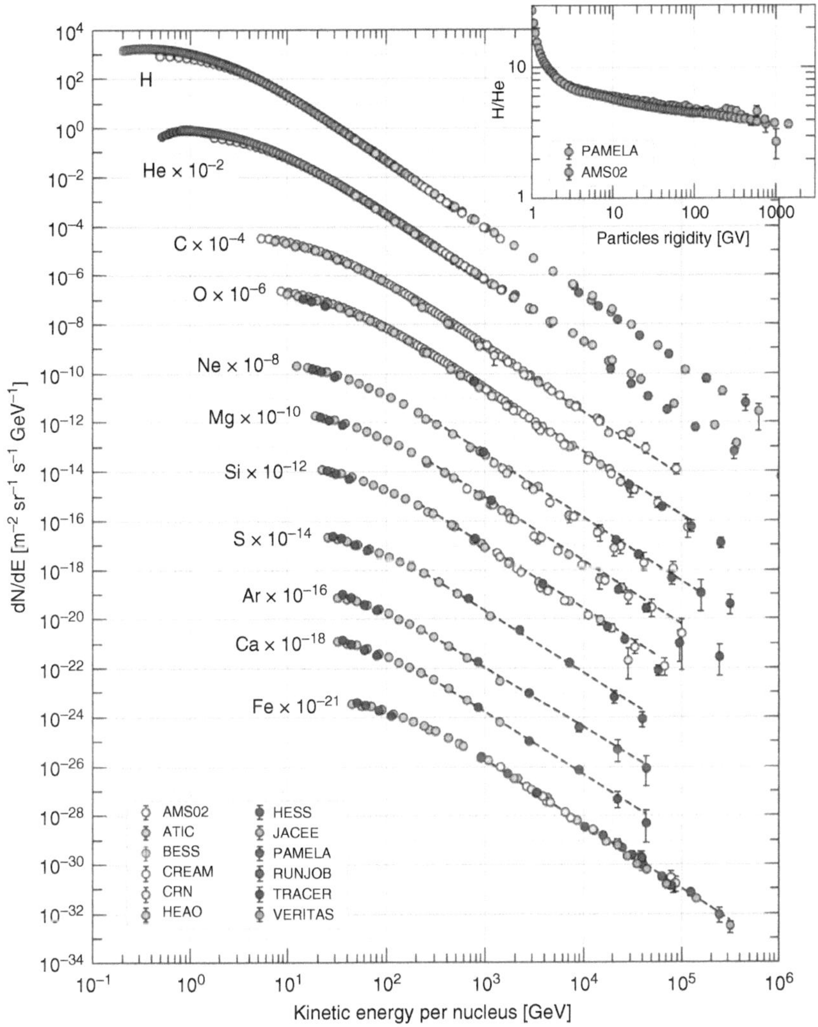

Fig. 1.9 ***Differential flux of galactic cosmic rays (GCR) (Particle Data Group, 2019).***

power-law distribution. The energy spectrum for shock-accelerated protons with kinetic energy of E is:

$$\frac{dN}{dE} \propto E^{-\gamma}, \tag{1.11}$$

where N is number of particles, and γ is the power-law spectral index expressed as:

$$\gamma = 2\frac{1+M^{-2}}{1-M^{-2}}, \tag{1.12}$$

where M is the Mach number.

The Sun travels through many different galactic environments for billion years at ~8 kpc distance from the galactic center. The most extreme space climate condition in the history of the Earth and the Sun is the possible encounter with a nearby *supernova remnant.* Assuming the blast wave of a supernova as a symmetric Sedov-Tayler solution (Sedov, 1959; Landau and Lifshitz, 1987), the radius R of the supernova shock front has a self-similar solution $R \propto (E_0/n)^{1/5}\, t^{2/5}$, where E_0 is the initial energy, n is the surrounding medium density, and t is the age. Fields et al. (2008) have suggested that the total pressure, P_{SN}, can be obtained as follows:

$$p_{\mathrm{SN}} = 3.3\left(\frac{R}{10\,\mathrm{pc}}\right)^{-3}\left(\frac{E_0}{10^{51}\,\mathrm{erg}}\right)\mathrm{nPa}, \tag{1.13}$$

The pressure of a nearby supernova remnant is comparable to the dynamic pressure of the present-day solar wind at 1 AU, as shown in Eq. (1.7). Due to the pressure balance, the size of the heliosphere can therefore be as small as 1 AU, and it becomes possible to frequently observe the heliopause from Earth-orbiting satellites. If the solar system was in such a wrong place in our galaxy, there is a possibility of catastrophic consequences on life (Kataoka et al., 2013a, 2014).

1.2 Geomagnetic field

The Earth is a magnetized planet. Major cities such as Tokyo, Japan are located at low and middle latitudes, deep under the invisible magnetic shield against charged particles. The magnetic field becomes visible to human eyes as the rayed appearance of aurora when energetic electrons precipitate along the geomagnetic field into the polar upper atmosphere.

In the spherical coordinate system with the Z-axis along the magnetic moment, the dipole magnetic field components are

$$\begin{pmatrix} B_r \\ B_\lambda \\ B_\varphi \end{pmatrix} = \begin{pmatrix} 2ar^{-3}\sin\lambda \\ -ar^{-3}\cos\lambda \\ 0 \end{pmatrix}, \tag{1.14}$$

where a is the magnetic moment, and the latitude λ is $\pi/2-\theta$. The strength of the magnetic field is therefore proportional to a and r^{-3}. If one selects a magnetic field-line at the magnetic equator at distance $r = r_{\mathrm{eq}}$, moving along the magnetic field, the magnetic field strength B is:

$$B = \frac{a}{r_{\mathrm{eq}}^3}\frac{\sqrt{1+3\sin^2\lambda}}{\cos^6\lambda} \tag{1.15}$$

A complex geomagnetic field is represented by means of *Gauss coefficients* such as g_{mn} and h_{mn}. These are coefficients of associated Legendre's polynomials of P_n^m multiplied by sine or cosine of longitude, which expands the geomagnetic potential into series of terms of degree n and order m, as briefly introduced below.

The current-free situation is:

$$\mathbf{J} = \nabla \times \mathbf{B} = 0. \tag{1.16}$$

The magnetic field can be generally expressed by a scalar potential W because rot-grad becomes zero:

$$\mathbf{B} = -\nabla W \tag{1.17}$$

The divergence-free nature of the magnetic field then gives the solution of the Laplace equation in the spherical coordinates by

$$W = r_E \sum_{n=1}^{\infty} \sum_{m=0}^{n} \left(\frac{r_E}{r} \right)^{n+1} P_n^m (\cos\theta)(g_n^m \cos m\phi + h_n^m \sin m\phi), \tag{1.18}$$

where r_E is the Earth's radius. Thus, the first degree of $n = 1$ can be written down as:

$$W = \frac{r_E^3}{r^2} (g_1^0 \cos\theta + g_1^1 \sin\theta \cos\phi + h_1^1 \sin\theta \sin\phi) \tag{1.19}$$

This is identical with the simple dipole equation in the spherical coordinate system:

$$W = \frac{\mu_0}{4\pi} \frac{\mathbf{M} \cdot \mathbf{r}}{r^3}, \tag{1.20}$$

where $\boldsymbol{M}$ is the magnetic moment and μ_0 is the permeability in vacuum.

Using the Gauss coefficients, we can define better coordinates such as apex latitude or corrected geomagnetic latitude (Richmond, 1995) for studying auroras than simple dipole latitude or dip latitude. Fig. 1.10 shows examples of the auroral zone in different age, calculated by apex latitude (Kataoka and Nakano, 2021), based on the IGRF model (Alken et al., 2021) and the paleomagnetic field model CALS (Korte and Constable, 2011). The original data to reconstruct the paleomagnetism are obtained, for example, from residual magnetism of kiln traces (Hirooka, 1971).

The geomagnetic field shows complete reversals of polarity at a rate of 2 to 3 per million years (Jacobs, 1994). The *geomagnetic reversal* gives an extreme space climate condition for a polarity-flipping time scale of an order of ~1000 years. The latest magnetic reversal event was identified 0.77 million years ago at Chibanian sector (Suganuma et al., 2021), when the total strength of the geomagnetic field reduced to an order of magnitude smaller than that of the

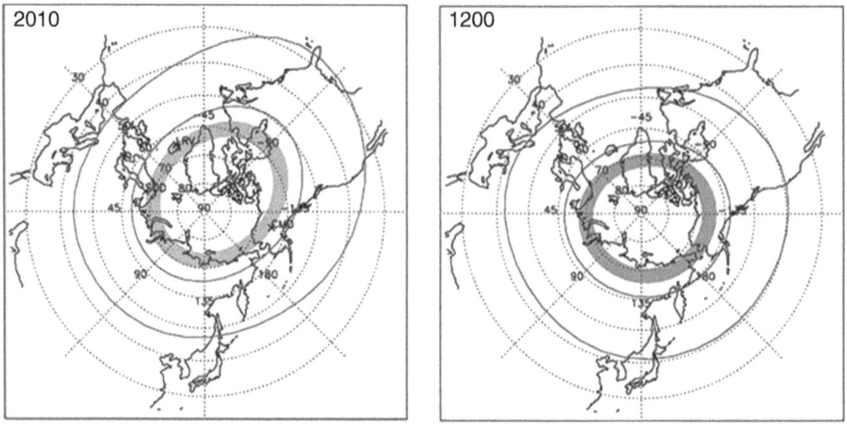

Fig. 1.10 ***Auroral zone (65°–70° magnetic latitude) with 60° and 40° magnetic latitudes in 2010 and 1200 AD (Kataoka and Nakano, 2021).***

current state. Magnetic reversal or magnetic excursion is a complex process related to the coupling between fluid convection in the outer core and dynamo action (Wicht and Tilgner, 2010). The possible consequences of magnetic reversals and excursions have been discussed by Glassmeier and Vogt (2010) and Cooper et al. (2021). As a possible effect of the magnetic moment on auroral ovals, 1.5 times weaker magnetic moment than the present value can expand the auroral oval toward lower latitudes (Ebihara and Tanaka, 2021).

The geomagnetic field prevents the intrusion of charged particles to the Earth. The gyro radius r_g of a particle with the mass m, charge q, speed v is given by:

$$r_g = \frac{\gamma m v_\perp}{qB}, \tag{1.21}$$

where the Lorentz factor $\gamma = (1 - v^2/c^2)^{-1/2}$, and c is the speed of light. For relativistic high-speed particles with $v \sim c$, the gyro radius of charged particles becomes independent of the kind of nuclides with different mass. The Loretnz factor is the ratio of kinetic energy and the rest energy (0.511 MeV for electron and 0.9315 GeV for proton) plus one. *Magnetic rigidity*, defined as Br_g is particle momentum per unit charge, characterizing the ability of cosmic rays to penetrate the geomagnetic field. For an ion with a mass number A (the number of neutrons and protons in the nucleus), charge Z (in units of proton charge), and kinetic energy E (in units of GeV/nucleon), the rigidity R is given by:

$$R = \frac{A}{Z}\sqrt{E^2 + 2M_0 E}\ \mathrm{GV}, \tag{1.22}$$

where $M_0 = 0.9315\ \mathrm{GeV}/c^2$ is the atomic mass unit.

Simplified Störmer's theory of gyrating particles in a dipole geomagnetic field (Walt, 1994) can be written as:

$$R_c = 14.9\cos^4 \lambda_c, \tag{1.23}$$

where all particles with a critical rigidity R_c reach at a point of geomagnetic latitude of $\lambda > \lambda_c$ (Störmer, 1955; Smart and Shea, 2005). Examples of the vertical cutoff rigidity map are shown in Fig. 1.11. For example, 100 MeV protons can reach the atmosphere of the Earth only at high latitude, while all of the >15 GeV protons or electrons can reach the equatorial atmosphere. The cutoff also depends on the geomagnetic

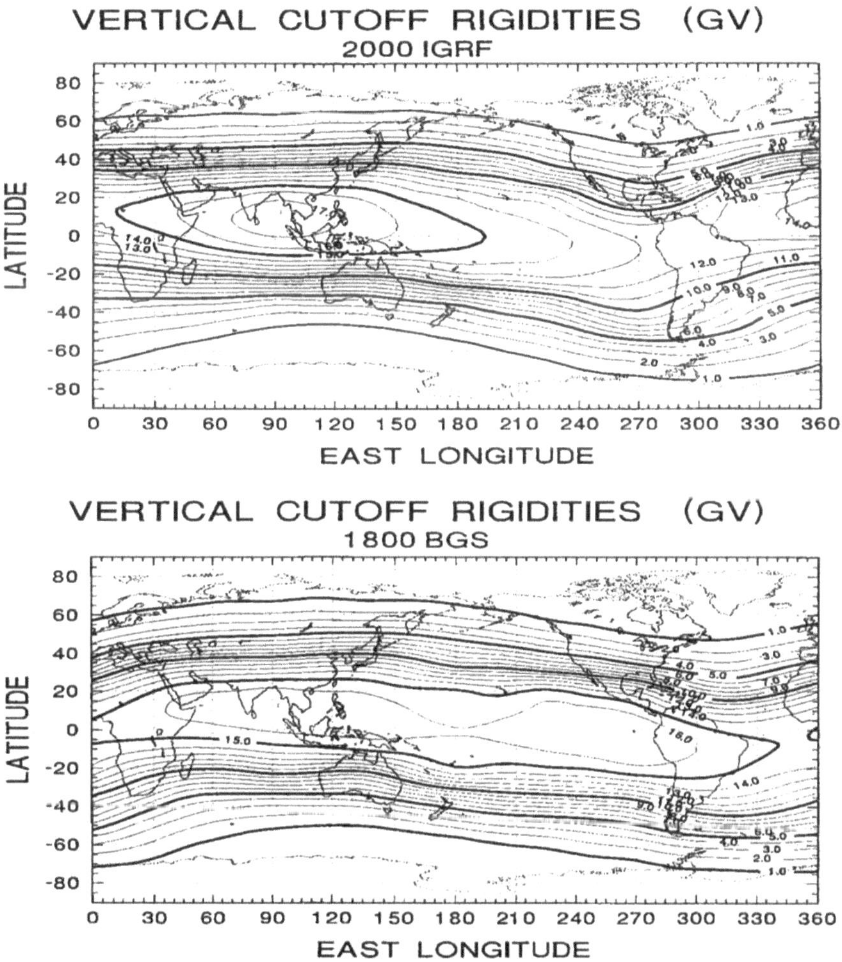

Fig. 1.11 ***Vertical cutoff rigidity contours in 2000 and 1800 AD (Smart and Shea, 2009).***

activity. Using the Dst index in nT, a hybrid model has been developed as follows:

$$R_c = 15.062\cos^4\left(\lambda_c + \mathrm{Dst}/19.11\right) - 0.363 \tag{1.24}$$

Ogliore et al. (2001) modified the model without a Dst dependence, and Leske et al. (2001) added the Dst correction. Comparing with these empirical models, O'Brien et al. (2018) demonstrated that low-altitude proton measurements are beneficial for monitoring the actual geomagnetic cutoff near real-time.

The Earth has been an electric planet for billion years due to the geomagnetic field and its action of the current generator against the solar wind. The shielded space by the electric current is called *magnetosphere*. The outer boundary or magnetosphere surface is termed *magnetopause*, and the surface current is called Chapman–Ferraro current (Chapman and Ferraro, 1931). From the cross product of **B** and momentum conservation Eq. (1.1), *diamagnetic current* and *inertial current* can be defined as follows:

$$\mathbf{J}_D = \frac{1}{B^2}\mathbf{B}\times\nabla p \tag{1.25}$$

$$\mathbf{J}_I = \frac{1}{B^2}\mathbf{B}\times\left(\rho\frac{\partial\mathbf{v}}{\partial t}\right) \tag{1.26}$$

The Chapman-Ferraro current and the tail current (Fig. 1.12) are diamagnetic currents that shield the magnetosphere, while the *field-aligned current* (FAC), flowing parallel to the magnetic field supplies the power of auroras and connect the magnetosphere and ionosphere (Fig. 1.13). The inertial current plays an essential role during transition, such as *substorms* and *sudden commencements* (SCs).

The geomagnetic field poses a magnetic pressure p_B against the dynamic pressure p_d of the solar wind plasma (Eq. 1.7). Assuming the typical magnetic field of 100 nT at *geosynchronous orbit* (GEO) at 6.6 R_E, also using the relationship $B \propto r^{-3}$ at equator, the magnetic pressure p_B at the magnetopause distance R_{MP} is:

$$p_B = \frac{B^2}{2\mu_0} = 0.33\left(\frac{B_0}{100\ \mathrm{nT}}\right)^2\left(\frac{R_{MP}}{10\ R_E}\right)^{-6}\ \mathrm{nPa}, \tag{1.27}$$

where μ_0 is the permeability in vacuum. During the extreme events, the magnetosphere size can shrink to 6 R_E, and the spacecraft at GEO

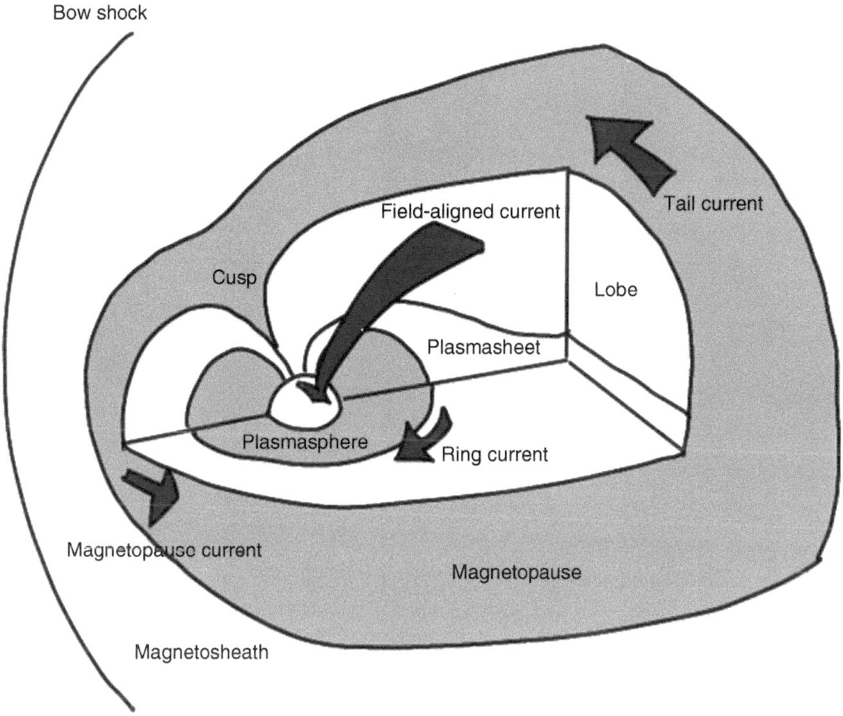

Fig. 1.12 ***Schematic illustration of the magnetosphere of the Earth.*** The downstream region of the bow shock is called magnetosheath, where the solar wind flow becomes subsonic. The surface current flows along the magnetopause. Hot plasma exists at cusp region in the dayside, and at plasma sheet region in the nightside. Void volume at the tail is called lobe region. Dense and cold plasma exist in the plasmasphere, where radiation belt electrons and ring current protons coexist.

experience *geosynchronous magnetopause crossing* (GMC). The closest magnetopause at 5.2 R_E was observed on 4 August, 1972 (Hoffman et al., 1975). The total accumulated duration of the GMC events is ~4 hours per year, and each duration of the GMC events is ranging from a few minutes to a few hours (Rufenach et al., 1989). Interestingly, the magnetopause remained close to geosynchronous orbit for more than 4 hours on 6 August, 1998 (Lugaz et al., 2015). Even the bow shock can be inside GEO, and such an extreme event occurred on 21 January, 2005, when the solar wind dynamic pressure exceeded 100 nPa (Dmitriev et al., 2014).

The magnetosphere consists of discontinuities separating different kinds of plasma (Fig. 1.12). The forefront of the discontinuity is the standing shock called *bow shock*, where the supersonic solar wind plasma becomes

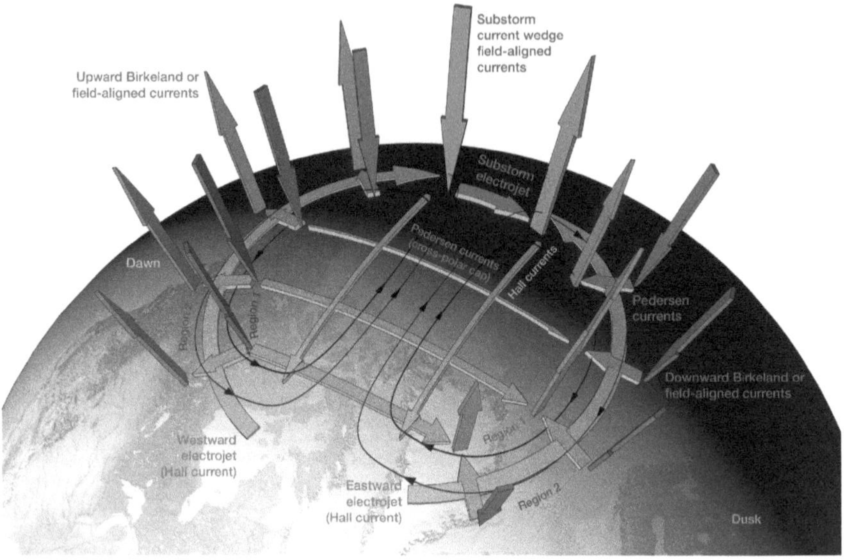

Fig. 1.13 ***Field-aligned currents (FAC) and other ionospheric currents (Palmroth et al., 2021).*** The pioneering work was given by Iijima and Potemura (1978).

subsonic. Plasma and magnetic fields are compressed at the bow shock following the Rankine–Hugoniot relationship (Szabo, 1994).

In the inner magnetosphere, that is, inside the geosynchronous orbit, different plasma populations coexist at the same location due to their collision-less nature. Trapped particles of the most energetic ions and electrons (~MeV) form the *radiation belts* (Van Allen and Frank, 1959). Ring current particles consist of less energetic ions and electrons (~sub-MeV), which are sufficiently large number to carry the large electric current that can deform the intrinsic magnetic field. Cold plasma of 1 to 10 eV flowing up from the atmosphere also coexists, corotating with the atmosphere, to form the *plasmasphere*. The plasmasphere is the upper-most part of the ionized atmosphere with the electron density of only 10^3 /cc, but it occupies the mass of the magnetosphere.

The solar wind flows against all planets in the solar system, and the magnetosphere is not unique for the Earth. Several extreme magnetospheres exist in the solar system. As shown in Table 1.1, the giant planets are strongly magnetized (Bagenal, 1992; Connerney et al., 1991, 2018) and their magnetospheres are huge (Russell, 2004).

In addition to the giant planets, Mercury has a weak magnetic moment (~4×10^{-4} of that of the Earth; Russell and Luhmann, 1997). Venus has no detectable intrinsic magnetic field (Russell, 1980), while Mars has a

Table 1.1 Magnetic field and basic parameters for the Earth and giant planets.

Planet	Earth	Jupiter	Saturn	Uranus	Neptune
Distance (au)	1	5	9–10	18–20	30
Radius (km)	6378	71,492	60,330	25,600	24,765
Spin period (hours)	24	9.9	10.2	17.2	16.1
Equatorial field (G)	0.31	4.2	0.21	0.14	0.23
Dipole tilt (deg)	11.3	−9.6	0	−59	−47

weak localized magnetic field from the magnetization of the Martian crust (Connerney et al., 2001), as also introduced in Chapter 5.

1.3 Atmosphere

Aurora is the atmospheric end-product of the violent energy flow that originates in the Sun. The visible auroras are emissions in the *ionosphere*, the uppermost atmospheric layer of the Earth, a region at 85 to 600 km altitude. At least two different aurora types exist. Discrete auroras (Fig. 1.14) are caused by beamed electrons accelerated by the potential drop along or parallel to the magnetic field. The formation mechanism of the auroral parallel electric field is an unsolved issue. Diffuse auroras are caused by pitch-angle scattering of energetic electrons trapped in the magnetosphere. There is another type of auroras known as *stable auroral red* (SAR) arc, which is caused by Coulomb collisions between ring current heavy ions and thermal electrons of the plasmasphere (Kozyra et al., 1987; Shiokawa et al., 2001). The SAR arcs are observed at equatorward region from the discrete or diffuse auroras, and occupy a part of so-called low-latitude red auroras (Shiokawa et al., 2005, 2013). The emission altitudes of discrete and diffuse auroras are approximately >100 km and <100 km, respectively (Fig. 1.15).

The energy spectrum of discrete aurora can be expressed by a Gaussian energy spectrum with characteristic energy E_c of a few keV and a Gaussian width W (Strickland et al., 1993):

$$f_{discrete}(E) = \frac{Q}{\pi^{3/2} W E_c} \exp\left(-\left(\frac{E - E_c}{W} \right)^2 \right), \quad (1.28)$$

where Q denotes the total energy flux. On the other hand, diffuse auroras typically have a hard spectrum, and an energy spectrum of a kappa

Fig. 1.14 ***Drawing of an aurora by Fridtjof Nansen (In Northern Mists, 1911).***

distribution (Frahm et al., 1997) can be assumed with a peak energy E_p of 1 keV and a kappa parameter κ, considering a high-energy tail in addition to Maxwell distribution:

$$f_{diffuse}(E) = \frac{Q}{2\pi E_p^3} E \left(1 + \left(\frac{E}{\kappa E_p} \right)^{-\kappa-1} \right) \tag{1.29}$$

The examples of these models and the resultant atmospheric ionization profiles are shown in Fig. 1.16.

The colors of auroras provide the information on the photo-chemical reactions occurring in the atmosphere. As an easily remembered relationship, the energy of a photon in eV is equal to 12,345 divided by

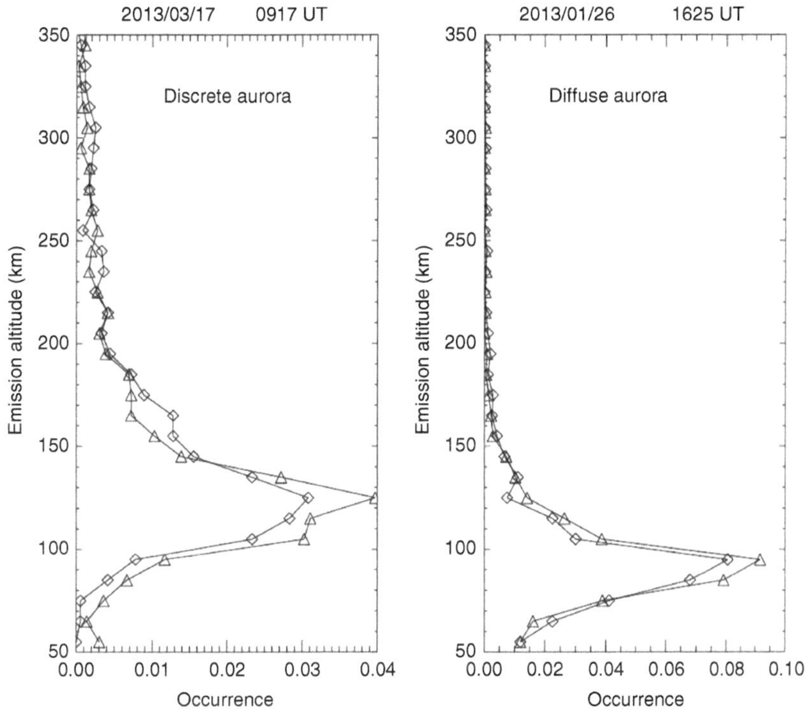

Fig. 1.15 ***Altitude profiles of (left) discrete and (right) diffuse auroras, as estimated from the stereoscopy using twin all-sky digital cameras located at 8-km apart (Kataoka et al., 2013b).***

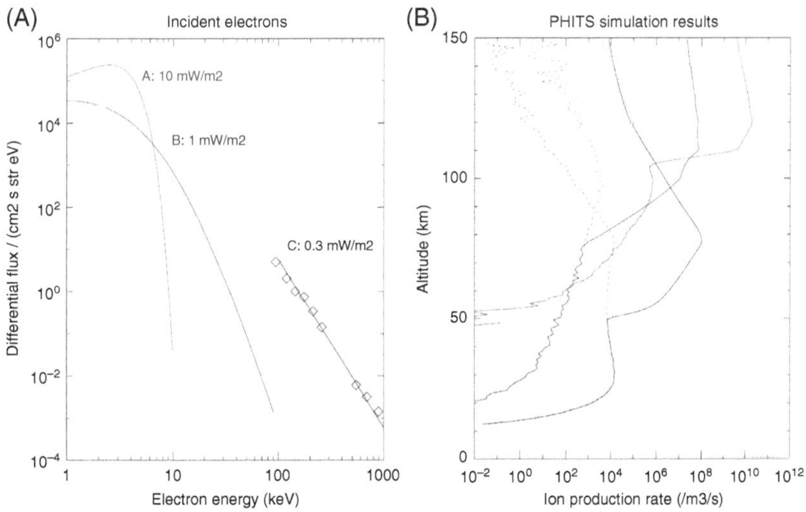

Fig. 1.16 ***Model energy spectra of (A) discrete aurora, (B) diffuse aurora, and (C) energetic electrons and those resultant atmospheric ionization profiles in the right panel (Kataoka et al., 2019).***

the wavelength in Angstroms. Thus, for example, photons of wavelength 6300 Å (auroral red line) and 5577 Å (auroral green line) have energy of 1.96 eV and 2.21 eV, respectively. An oxygen atom can accept exactly 1.96 eV, the energy required to raise it to the first excited state. Alternatively, the oxygen atom can accept 4.17 eV, thus rising to its second excited state.

Although the primary energy to ionize the atmosphere is provided by precipitating electrons, only a small part of the auroral emission is caused by the direct collisional excitation through the precipitating electrons or their secondaries. The major part is released through complex chemical reactions in the weakely ionized atmosphere. For example, the predominant green aurora is caused by the following reactions (Rees, 1984):

$$N^+ + O_2 \rightarrow NO^+ + O(^1S) \tag{1.30}$$

The excited oxygen atom then transits to a lower excitation state by emitting a photon.

$$O(^1S) \rightarrow O(^1D) + \text{photon}\,(557.7\,\text{nm}) \tag{1.31}$$

The red line is then emitted when this metastable state goes to the ground state.

$$O(^1D) \rightarrow O(^3P) + \text{photon}\,(630.0\,\text{nm}) \tag{1.32}$$

The $O(^1D)$ state can also be directly excited through dissociative recombination.

$$O_2^+ + e \rightarrow O + O(^1D) \tag{1.33}$$

Nitrogen molecules can also be ionized and excited by primary electrons:

$$N_2 + e_{\text{primary}} \rightarrow N_2^+\left(B^2\,\Sigma_u^+\right) + e_{\text{primary}} + e_{\text{secondary}} \tag{1.34}$$

This excited state of the ionized molecule is a vibrational state. During the transition to other vibration states, a whole band of colors in the blue-violet range is emitted. Further, emissions of neutral nitrogen molecule fall within the red and ultraviolet bands (Rees, 1989). An example of auroral spectra is shown in Fig. 1.17.

The ionosphere is defined by enhanced electron density layers (Fig. 1.18). Daytime X-rays and ultraviolet (UV) rays from the Sun primary ionize the global atmosphere. The E-region at ~100 km disappears during night because the dissociative recombination

$$NO^+ + e \rightarrow N + O \tag{1.35}$$

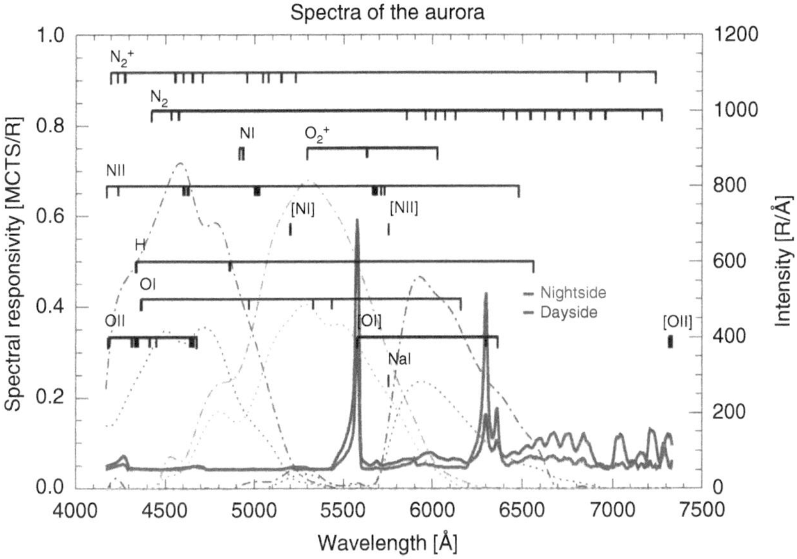

Fig. 1.17 ***The spectra of auroras with atomic lines and molecular bands.*** The sensitivities of Nikon cameras are also shown by dotted curves (after Sigernes et al., 2008).

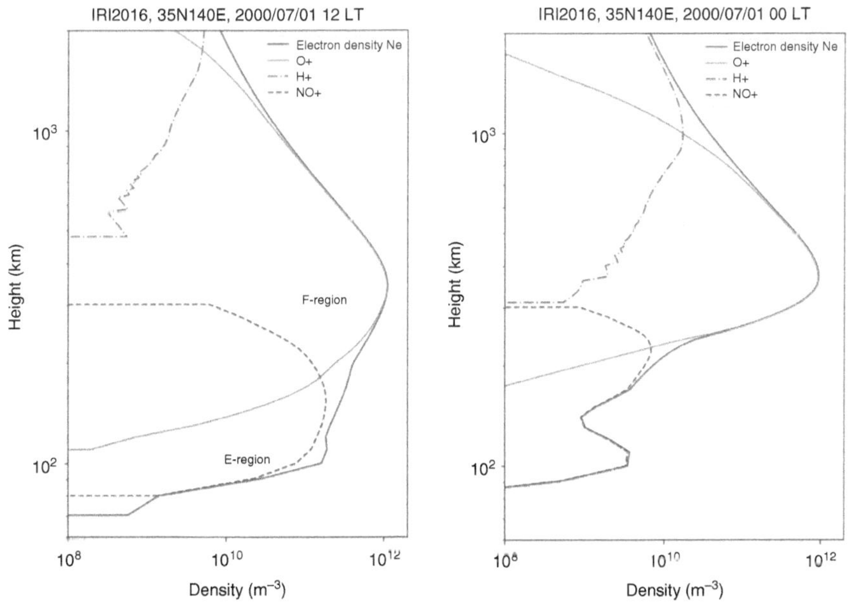

Fig. 1.18 ***Altitude profiles of electron density and the dominant ions (IRI 2016 model) during (left) day time and (right) night time.*** The electron density peak at 200 to 300 km is called F-region, while the peak at 100 km is called E-region.

is fast. The F-region at >200 km lasts long because the recombination of

$$O^+ + O_2 \rightarrow O_2^+ + O \tag{1.36}$$

$$O_2^+ + e \rightarrow O + O \tag{1.37}$$

is relatively slow due to the slow reaction of Eq. (1.36). These different time scales originate in the molecular ion composition in the E-region and atomic ion composition in the F-region.

The ionospheric density is usually represented by the *total electron content* (TEC, height integrated value) or NmF2 (peak value of electron density). In addition to the predictable daily variation associated with daylight ionizations and recombination, unusual ionospheric density enhancement or depletion are occasionally observed, which are called *positive ionospheric storms* and *negative ionospheric storms*, respectively.

At high latitude, the ionization source is dominated by auroral electrons in the night sky. The large number of keV electrons carry the FAC of an order of million Ampere along the magnetic field. The current circuit satisfies the Kirchhoff's law to couple the ionosphere and the magnetosphere.

Weakly ionized plasma of the ionosphere is anisotropic because the geomagnetic field is strong and electric currents do not always flow parallel to the direction of the electric field. Let us assume that the Z component directs vertically along the ambient magnetic field, X component directs toward the electric field, and Y component directs perpendicular to the electric field and magnetic field. Combining the steady-state momentum equations of electrons and ions, the electric current can be expressed as:

$$\mathbf{j} = \tilde{\sigma}\left(\mathbf{E} + \mathbf{U} \times \mathbf{B}\right), \tag{1.38}$$

where **U** is the neutral gas velocity. The conductivity tensor $\tilde{\sigma}$ is introduced as:

$$\tilde{\sigma} = \begin{pmatrix} \sigma_P & -\sigma_H & 0 \\ \sigma_H & \sigma_P & 0 \\ 0 & 0 & \sigma_\parallel \end{pmatrix}, \tag{1.39}$$

where *Pedersen and Hall conductivities* are defined as follows:

$$\sigma_P = \frac{eN_e}{B}\left(\frac{\omega_e \nu_{en}}{\omega_e^2 + \nu_{en}^2} + \frac{\omega_i \nu_{in}}{\omega_i^2 + \nu_{in}^2}\right) \tag{1.40}$$

$$\sigma_H = \frac{eN_e}{B}\left(\frac{\omega_e^2}{\omega_e^2 + \nu_{en}^2} + \frac{\omega_i^2}{\omega_i^2 + \nu_{in}^2}\right) \tag{1.41}$$

$$\sigma_{\parallel} = e^2 N_e \left(\frac{1}{m_e(\nu_{en} + \nu_{in})} + \frac{1}{m_i \nu_{in}} \right) \tag{1.42}$$

These currents depends on the collision frequencies ν and gyro frequencies ω. The resultant current system in the ionosphere is three-dimensional (Fig. 1.13). Roughly speaking, Hall current dominates in the E-region, while Pederson current dominates in the F-region.

Fukushima's theory (Fukushima, 1969, 1976) relates the horizontal magnetic field deviation at ground level to the Hall current in the ionosphere. This is based on the Ampere's law that cancels out the contributions from Pedersen current and field-aligned current. High-latitude ground-based magnetometer data are therefore used to estimate the Hall current pattern, and this is called *equivalent current*. Ground-based HF radars directly measure the Doppler velocity of horizontal plasma motion, which is in opposite direction of the Hall current. The horizontal velocity $\mathbf{v}$ of the ionospheric plasma is a drift velocity given by:

$$\mathbf{v} = \frac{\mathbf{E} \times \mathbf{B}}{B^2} \tag{1.43}$$

Note here that the *drift velocity* perpendicular to the external force $\mathbf{F}$ can be generalized as

$$\mathbf{v} = \frac{\mathbf{F} \times \mathbf{B}}{qB^2} \tag{1.44}$$

The *ionospheric convection* can therefore be approximated as incompressible two-dimensional fluid motion with an electrostatic potential as the streamline function. Such an electrostatic property of the global electric field distribution in the ionosphere was first reported by Kikuchi et al. (1978). The theory of the horizontal electromagnetic wave transmission powered by the FAC has been fully tested by ground-based observation networks (Kikuchi et al., 2021). The electric potential map in the polar region therefore works as the "space weather map" of the ionosphere, which is the space version of the pressure pattern in the weather forecast. Global HF radar network SuperDARN has successfully constructed the space weather map (Fig. 1.19). The history of the SuperDARN project was introduced by Greenwald (2021).

We are living in the atmosphere of the Earth, which has the *middle atmosphere*, formed by UV-ray absorption by ozone. It is hard to protect life from the UV-ray damages without the ozone layer (Marshall et al., 2020). The ozone layer is vulnerable to space weather, as discussed in Section 2.3.

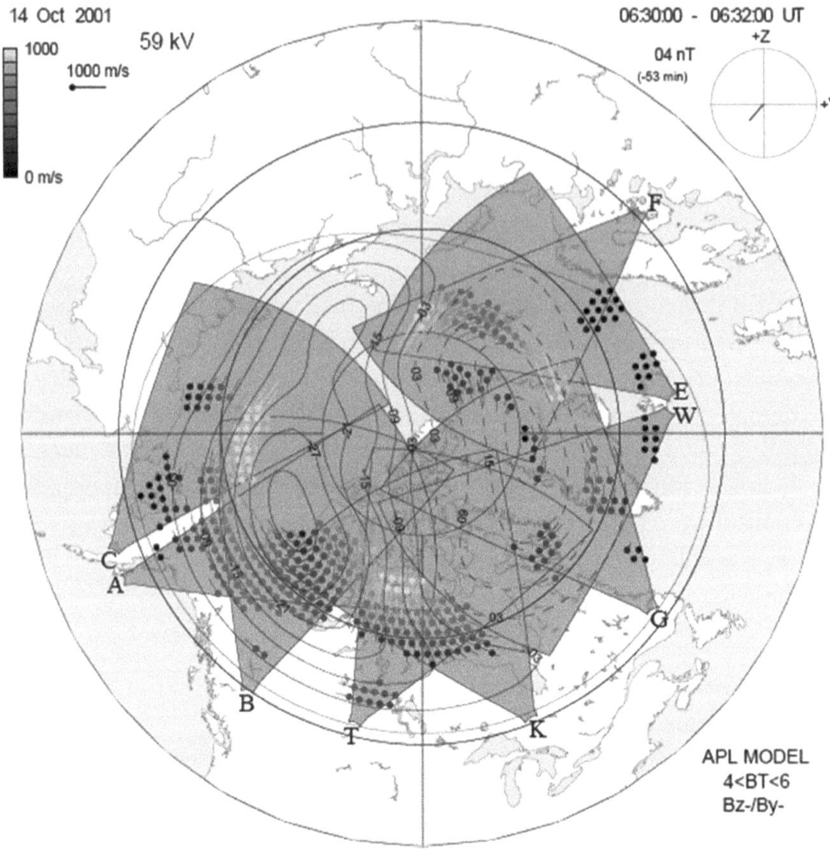

Fig. 1.19 ***Ionospheric convection pattern as derived from SuperDARN (Hoque et al., 2021).*** The pioneering work was given by Ruohoniemi et al. (1998).

The upper part from the temperature peak in the middle atmosphere is called *mesosphere*, and the lower part is called *stratosphere* (Fig. 1.20). The upper atmosphere, or the *thermosphere*, overlaps with the ionosphere at >80 km. In the thermosphere, the atmosphere changes from being a homogeneous mixture primarily comprising of molecular nitrogen to a diffusively separated gas mixture dominated by atomic oxygen. Finally, the lower atmosphere is called the *troposphere*.

The atmospheric shelter against cosmic rays weakens when we fly high. From the Eq. (1.1), the hydrostatic equilibrium leads to:

$$\rho = \rho_0 \exp\left(\frac{-(z - z_0)}{H}\right), \tag{1.45}$$

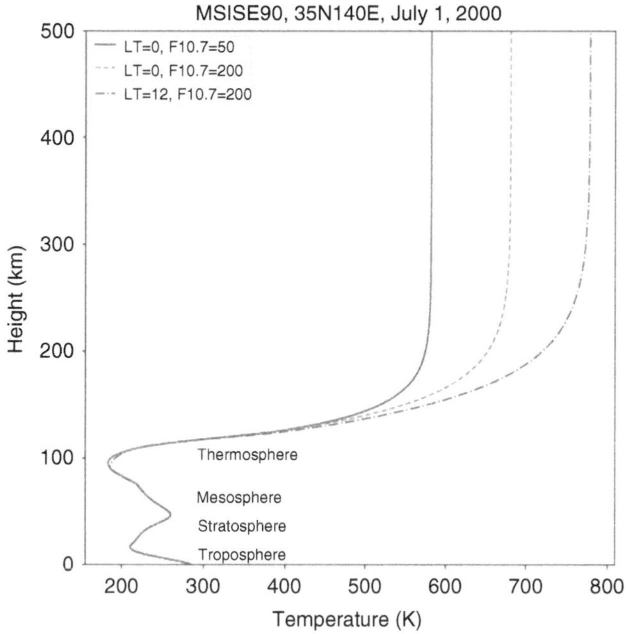

Fig. 1.20 ***Altitude profiles of the atmospheric temperature for different solar activity and different local time (MSISE 90 model).***

where $H = k_B T/mg$ is the scale height, with Boltzmann constant k_B, atmospheric temperature T, mean molecular wight m, and gravitational acceleration unit g. The exponential decrease of the atmospheric density ρ significantly influences the effective dose rates at ground level and flight altitude, as the GCR dose rate is hundred and thousand times higher at ~12 km flight altitude and ISS at ~400 km altitude, respectively. ISS is therefore an extreme situation because it sits just above the atmospheric shelter, although it is still behind the geomagnetic shield.

The troposphere is always ionized by GCR (Fig. 1.21). Nuclear spallation reactions among cosmic rays and atmospheric particles cause *air shower* (Fig. 1.22). The air shower contains penetrable populations such as muons and neutrons which contribute to atmospheric ionization as well as radiation exposure of humans in the airplane. The typical effective dose at ground is ~2 mSv per year. During the encounter with nearby supernova remnants, the effective dose rate increases to ~10^4 times in the troposphere (Kataoka et al., 2014).

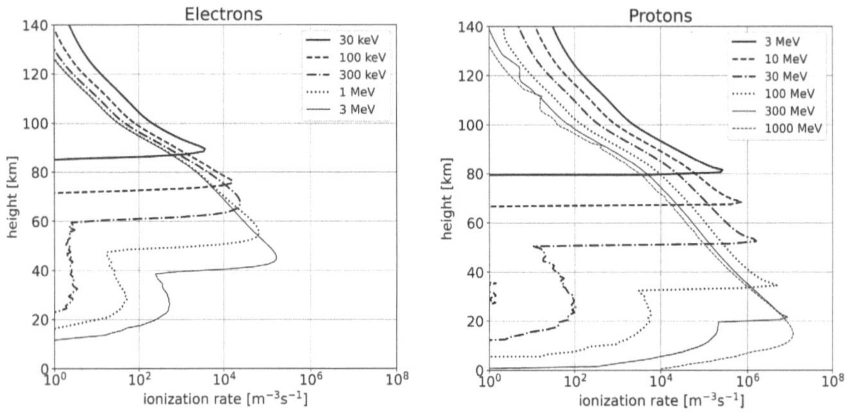

Fig. 1.21 ***Altitude profiles of the atmospheric ionization rate caused by precipitating electrons and protons, as calculated from PHITS.*** MSIS model is used as the density profile. The secondary peak at the ~30 km altitude in the left panel is due to electron Bremsstrahlung X-rays, while the secondary peak at ~20 km in the right panel is due to secondary electrons.

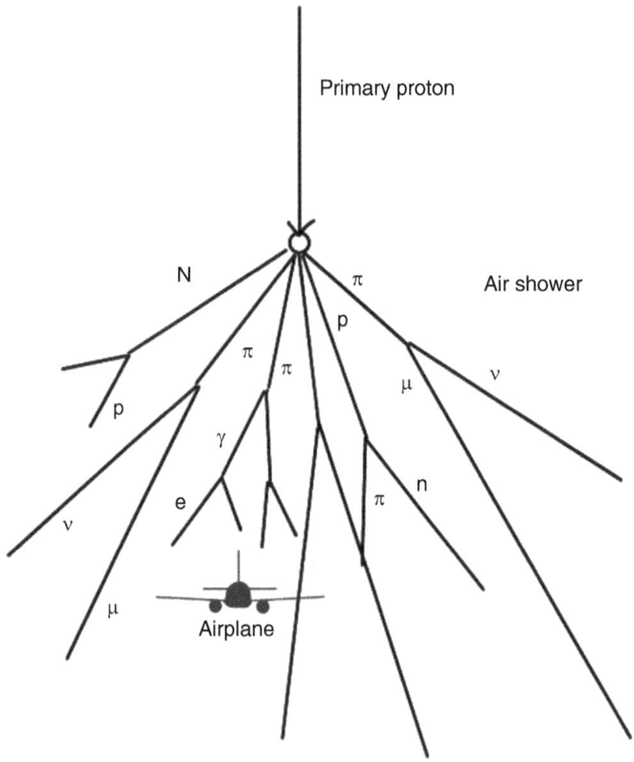

Fig. 1.22 ***Schematic diagram of the air shower caused by the collision between an energetic proton and atmospheric molecules.***

Constructing large commercial satellite constellations began in 2018 when SpaceX launched its first Starlink satellite prototypes. The potential addition of tens of thousands of objects to the low-Earth orbit (LEO) below 1000 km will escalate the risk of catastrophic and cascading collisions (Kessler et al., 2010). Atmospheric drag in LEO is the dominant source of errors in numerical modeling to predict the satellite position, which affects satellite mission planning, orbit and reentry prediction, and collision avoidance (Thayer et al., 2021).

The atmospheric drag force F is described by:

$$F = \frac{1}{2}\rho v^2 C_d A, \tag{1.46}$$

where v is the speed of spacecraft, A is cross section of the spacecraft in the traveling direction, and C_d is air drag coefficient determined by the shape of the spacecraft. Eqs. (1.45) and (1.46) give the essential difficulties of predicting the consequences from atmospheric drag. For example, when the satellite lowers the orbit by a small amount, the atmospheric drag enhances exponentially due to density increase and further lowers the orbit. Highly accurate and real-time observations of atmospheric density in the thermosphere are therefore necessary for safe reentry operations. In another example, atmospheric temperature change in decadal time scale matters. The thermospheric temperature depends on solar cycles (Fig. 1.20) as well as global climate change in the troposphere, both of which are difficult to predict till date (Lastovicka, 2021).

CHAPTER 2

Disturbed space weather

2.1 Solar flares

Solar flares are the aurora of the Sun. The plasma explosions in the solar corona go in two different directions, toward the Sun and away from the Sun. The former becomes flares, while the latter becomes coronal mass ejections (CME). The explosion is caused by the emergence of sunspots (Fig. 2.1) and the resultant reaction with the coronal magnetic field. The emerged sunspot magnetic field alters the ambient coronal magnetic field into different topology, as if coronal magnetic field meets a new magnetosphere. The magnetosphere of sunspots is called the *active region*, where the resultant magnetic field topology dynamically evolves. *Magnetic reconnection* is an intuitive concept to interpret the rapid change in the magnetic field topology leading to plasma explosion (Parker, 1957; Petschek, 1964; Sweet, 1958).

One of the theoretically predicted patterns from the magnetic reconnection is a cusp-like structure above coronal loop, which appears just after the reconnection (e.g., Yokoyama and Shibata, 1998; Takasao and Shibata, 2016). The typical solar flare is illustrated in Fig. 2.2, although the actual magnetic field structure is complex, and advanced concept such as null lines and null points has been discussed to understand the complex magnetic field topology (Pontin et al., 2013).

X-ray intensity has been used as the measure of the flare intensity. The flare activities, that is, the sudden increase of X-rays and UV for several hours, directly impacting atmospheric ionization, which leads to communication blackouts. An extreme case is known as *white light flare*, which was first observed via visible light on September 1, 1859, as shown in Fig. 2.3 (Carrington, 1860; Hodgson, 1960). At this moment people first noticed that the Sun was explosive to release intense electromagnetic waves. Carrington noted that "one swallow does not make a summer," even though he noticed that the worldwide auroral activity followed a day after the flare. In fact, the solar flares cannot directly cause auroras. CMEs, which can directly cause magnetic storms as introduced in Section 2.2, were then identified more than 100 years later.

Solar flare magnitude is categorized by alphabets (Fig. 2.4), according to the peak intensity I (W/m^2) between 1 and 8 Angstroms. The peak intensity of C, M, and X-class flares exceed 10^{-6}, 10^{-5}, and 10^{-4} (W/m^2), respectively.

Extreme Space Weather
DOI: https://doi.org/10.1016/B978-0-12-822537-0.00005-3

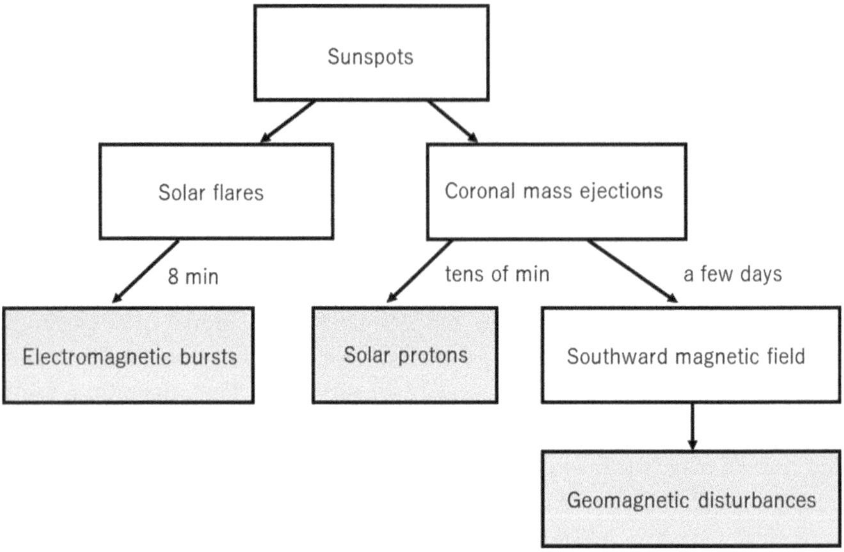

Fig. 2.1 ***Flow diagram indicating explosive energy release around sunspots to the disturbed space weather such as electromagnetic bursts, solar protons, and geomagnetic disturbances.***

The next category beyond X-class is the X10 class. The NOAA's operational category R1–R5 also exists as summarized in Table 2.1.

The Carrington flare is estimated to be ~X45 class (Cliver and Dietrich, 2013), based on the characteristic waveform of the ground-based mag-

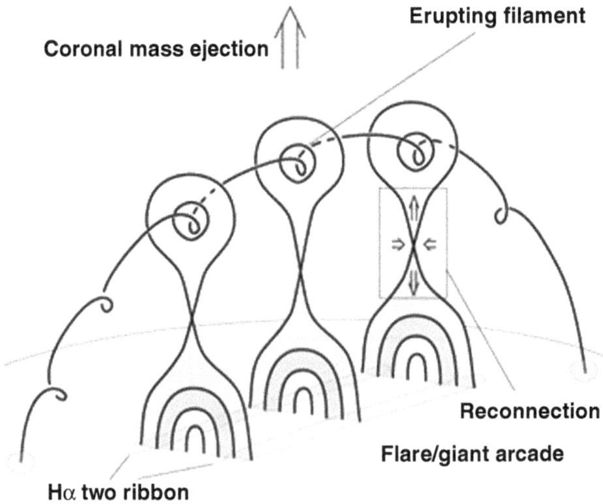

Fig. 2.2 ***Schematics illustration of a solar flare, reconnection, and erupting filament (Shiota, 2005): Courtesy of D. Shiota.***

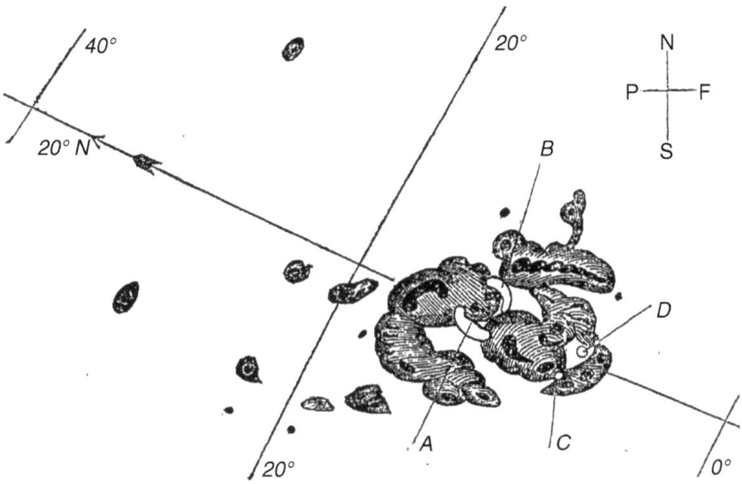

Fig. 2.3 The sketch of a complex sunspot on September 1, 1859, with bright flares (A–D), on September 1, 1859 (Carrington, 1859).

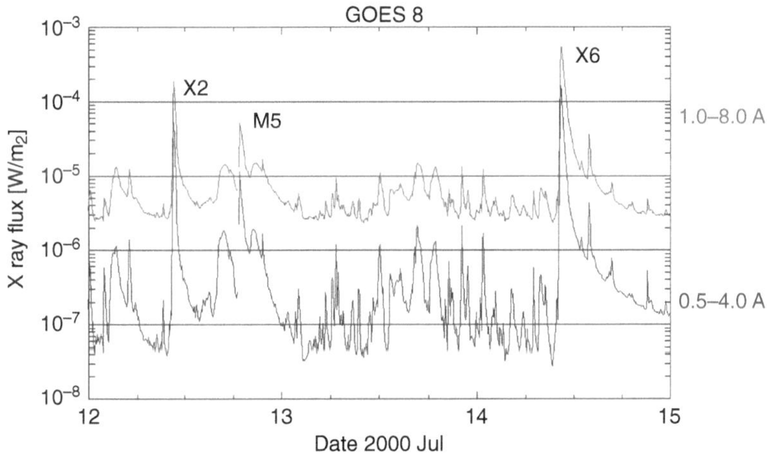

Fig. 2.4 ***So-called light curves of solar flares, which is used to define flare magnitude.*** These data were obtained real-time by GOES spacecraft during the Bastille day event. Everyone involved in the space weather is carefully watching this monitoring data as the first step.

Table 2.1 R-scale of NOAA/SWPC to categorize the impact of radio blackout impacts.

R-scale	R1	R2	R3	R4	R5
Description	Minor	Moderate	Strong	Severe	Extreme
Physical measure: X-ray	M1	M5	X1	X10	X20
Average frequency in 11 years	2000	350	175	8	Less than 1

netometer record called *solar flare effect*, which is caused by a transient enhancement of the ionospheric conductivity. Carrington flare amplitude is comparable with that of the largest observed ~X40 flares (X28 was observed at limb) on November 4, 2003 (Brodrick et al., 2005). The probability of super flare occurrence (beyond the X100 class) on the Sun is estimated to be one in 10^3–10^4 years (Shibata et al., 2013). Super flares have been observed in other stars (Schaefer, 2000). Maehara et al. (2012, 2015) identified a number of X1000–X10000 flares (10^{33}–10^{36} erg), while Notsu et al. (2019) showed the likely upper limit of 5×10^{34} erg for slowly rotating stars like the Sun. The possible consequences of super flares on exoplanets are actively discussed along the line of the possibility to host a life (Garraffo et al., 2016). What happens if the Earth received the largest imaginable super flare? White-light flares on the Sun change the brightness by an order of 0.01%, and such strong flares have a visible-light energy of an order of 10^{32} erg. The largest super flare of 10^{36} erg can therefore roughly double the solar constant. The equilibrium temperature increases by ~20%, which implies that the dayside Earth is literally cooked for a few hours.

These X-rays are dominantly emitted from the heated electrons of $>10^7$ K in the coronal loop. *Solar radio bursts* are other electromagnetic waves emitted from accelerated electrons around the flare site, which produce the direct noise on radio receivers. An example of the dynamic spectrum of solar radio burst is schematically depicted in Fig. 2.5. Frequently used in space weather are Type II and Type III. From type II radio burst, we can roughly estimate the speed of propagating shocks, because we can convert the radio frequency into distance using electron density model. Type III is a bookmark when flare occurs and evidences the signature of escaping electrons leaving the surface along open field lines.

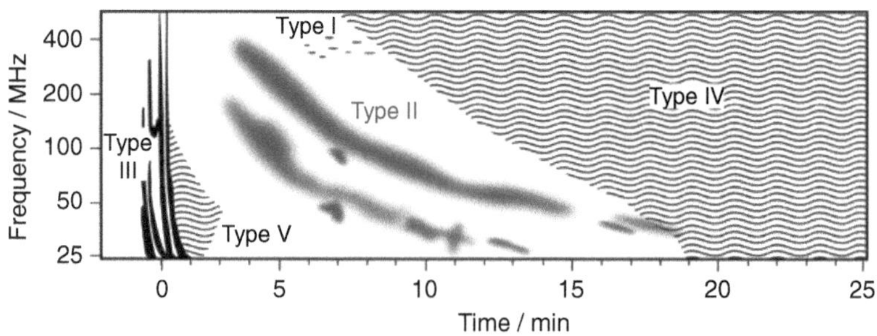

Fig. 2.5 ***Schematic diagram of different types of solar radio bursts categorized by the dynamic spectrum features.*** The zero time is the timing of solar flare (Ganse et al., 2012).

2.2 Coronal mass ejections

CMEs are the traveling magnetosphere departed from the Sun to the interplanetary space, that is, a foreign body of magnetized plasma ejected into the ambient solar wind. Fast CME forms *interplanetary shock* ahead of the CME. The shock downstream is called sheath region. Before the space age, Gold (1955) predicted the existence of high-speed plasma ejected from the Sun to form the shock ahead, which abruptly compresses the magnetosphere, in a process called SC. The Earth-directed CME appears ring-shaped in the coronagraph placed along the Sun-Earth line and is known as halo CME. Howard et al. (1982) has associated the halo CMEs to magnetic storm causes. In fact, the major magnetic storms are caused by the twisted magnetic field embedded in such halo CMEs as well as the compressed magnetic field in the sheath region (Kataoka et al., 2005a).

During solar eclipse, *red prominences* can be seen by the naked eye at the solar limb. When they are on the solar disk, the same phenomena are called *dark filaments* because they are the dark regions as observed by Hα line at 656.28 nm. The filaments are the cool and dense chromospheric plasma, sustained by the coronal magnetic field (Kaneko and Yokoyama, 2017). CMEs sometimes correspond to *disappearing filament* or *filament eruption* (Fig. 2.2). The magnetic polarity or helicity of the filaments therefore provide important hint to forecast the magnetic field structure in CMEs (Fig. 2.6).

After long-duration flares, the occurrence of *coronal dimming* in X-ray and Extreme UV (EUV) and/or large-scale coronal loop formation called *giant arcades* are known as the counterpart of CMEs (Harrison et al., 2003; Shiota et al., 2005). However, in some cases, any solar counterpart of CMEs in low corona (flares, dimming, filament eruptions) is missing, that is, no trace left behind the CMEs (Robbrecht et al., 2009). Those "stealth CMEs" have been intensively studied (Nitta and Mulligan, 2017; Palmerio et al., 2021).

Interplanetary CMEs have been identified by the strong magnetic field in the in-situ observation data of the solar wind (Fig. 2.7), and therefore called *magnetic clouds*. Other features include slowly rotating magnetic field, the so-called *flux rope* structure, and low plasma temperature due to the adiabatic expansion of the flx rope (Burlaga et al., 2001; Lopez, 1987; Richardson and Cane, 1995). Rotation polarity or magnetic helicity should reflect the solar magnetic field (Fig. 2.6). Clarifying the magnetic field directions is important for forecasting space weather, because the CME with southward-directing magnetic field (SBZ) drives major substorms and magnetic storms (Kilpua et al., 2017), as introduced in Section 2.4.

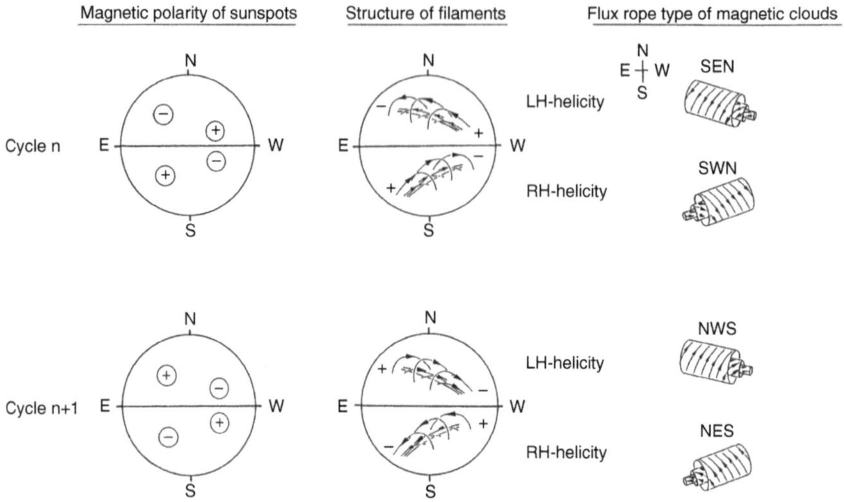

Fig. 2.6 ***Schematic diagrams showing the association between sunspots, filaments, and magnetic clouds (coronal mass ejections identified at interplanetary space).*** South-to-north rotating CMEs are dominant during odd (n) cycles, whereas north-to-south rotating CMEs dominate during even (n+1) cycles (Bothmer and Rust, 1997).

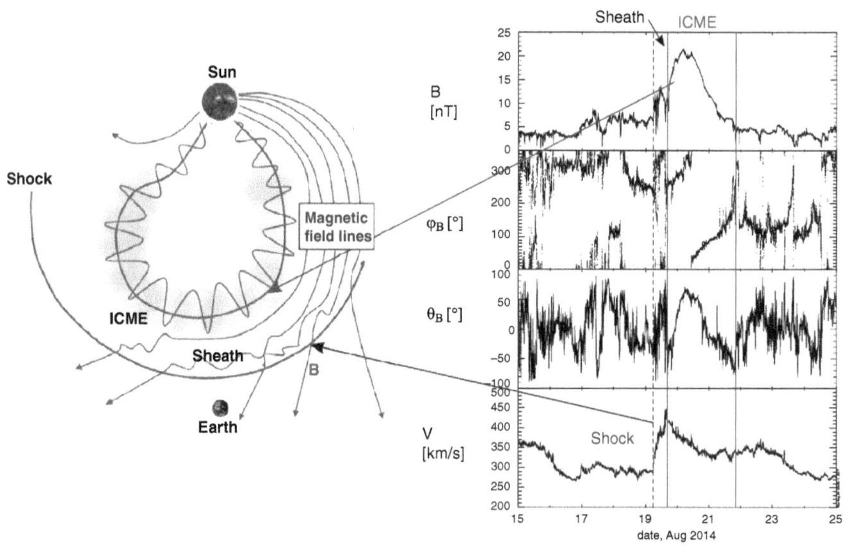

Fig. 2.7 ***Schematic diagram showing a coronal mass ejection (CME; left) and a sample data across the CME (right).*** From top to bottom: magnetic field strength, azimuth angle of the magnetic field from the Sun–Earth line, the north–south angle in the GSM coordinate system, and the solar wind speed (Kilpua et al., 2017).

Coronagraph was the game-changing technology for the space weather forecast. The NASA's OSO-7 mission first obtained direct coronagraph images that observed the CMEs in 1971 (Tousey, 1973). It then took a few decades for many researchers to recognize the essential role of CMEs in space weather. In 1993, the paper entitled "The solar flare myth" was published (Gosling, 1993), which even argued that solar flares do *not* generally play a central role in producing major transient disturbances in geomagnetic field. The history of CMEs and space weather is summarized by Gopalswamy (2016). Gopalswamy et al. (2005) presented a good example of the clear relationship between CMEs and magnetic storms (Fig. 2.8).

The average speed of CMEs are 300 and 500 km/s in the solar minimum and maximum, respectively (Yashiro et al., 2004). The average speed of halo CMEs (~1000 km/s) is about twice that of normal CMEs (~430 km/s). Extremely fast CMEs of >1500 km/s are rare, occupying ~0.5% of all CMEs (Wang and Zhang, 2007). The high speed of halo CMEs is not only from the sky-plane projection effects but also from the fact that they are fierce. The CME mass is in the rage from 10^{13} to 10^{16} g with a peak value at 3.4×10^{14} g (Vourlidas et al., 2010), which is comparable to the mass of comets. The kinetic energy is between 10^{27} to 10^{32} erg

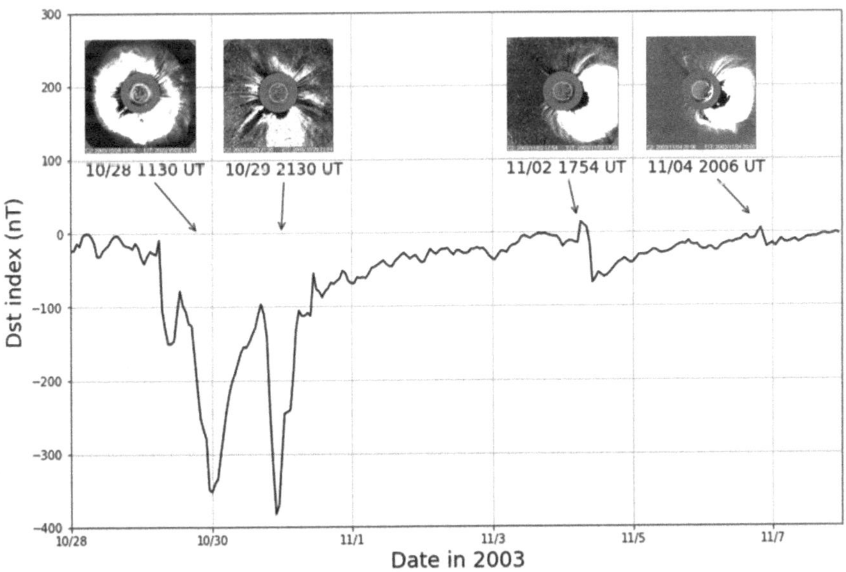

Fig. 2.8 ***Halo coronal mass ejections (CMEs) causing large magnetic storms and sudden commencements (SCs) in October/November, 2003.*** See also Gopalswamy et al. (2005).

with a peak value at 8.5×10^{29} erg (Vourlidas et al., 2010), which is comparable to the energy release from the solar flare. The fastest CME was 3675 km/s on January 20, 2005 (Gopalswamy et al., 2012).

CMEs are the main cause of extreme solar wind parameters at 1 AU. Cliver et al. (1990) discussed that the maximum solar wind speed at 1 AU would be ~2000 km/s, which was observed on August 4, 1972. In fact, the solar wind speed exceeded 2000 km/s during the October-November 2003 storm (Gopalswamy, 2005). On July 23, 2012, the maximum solar wind speed was 2250 km/s at the shock, while the maximum magnetic field strength was 109 nT at magnetic cloud (Russell et al., 2013). The maximum density of ~200/cc may be reasonable as the extreme value (Araki, 2014). The possible set of extreme solar wind parameters causing giant magnetic storms is therefore not randomly selected by a slot-machine manner, and it is determined by the interaction among the multiple CMEs and ambient solar wind structure (Kataoka et al., 2015a).

Carrington event is not the fastest-transit event. On August 4, 1972, the transit time from flare occurrence (0621 UT) to shock arrival (2054 UT) at 1 AU was only 14.6 hours (Vaisberg and Zastenker, 1976). The second fastest event is the Carrington event on September 1, 1859, with a 17.6 hour transit time (Cliver and Svalgaard, 2004). On July 23, 2012, the transit time from the onset (0208 UT) to shock arrival (1630 UT) at STEREO satellite at 0.96 AU was only 14.4 hours (Russell et al., 2013). The extreme interplanetary shock on July 23, 2012 was interesting because the pressure of energetic particles exceeded gas pressure and magnetic pressure (Russell et al., 2013; Lario et al., 2015). This is an extreme form of shocks beyond the Rankine–Hugoniot relation, known as *cosmic-ray modified shocks* (Drury and Voelk, 1981; Terasawa et al., 1998).

Extremely weak solar wind conditions also exist associated with CMEs. In the trailing edge of fast CMEs, both the solar wind density and magnetic field tend to be very small due to the solar wind rarefaction. Kataoka and Miyoshi (2008) discussed that such situation causes unusual inflation of the magnetosphere which can extremely enhance the outer belt electron flux. As a different type of extremely weak solar wind condition, May 10-12, 1999 is known as "the day solar wind disappeared." Many unusual space weather phenomena were observed, associated with the etremely low density of the solar wind (Anderson et al., 2000; Parks et al., 2000; Weatherwax et al., 2000; Nishitani et al., 2003).

2.3 Solar energetic particles

The storm time of *solar energetic particles* (SEPs) is called *solar proton event* because the SEP is dominantly composed of a proton. The solar proton events are monitored by the increase of 10 MeV protons, and the identification criterion is 10^3 PFU. There are two different types of SEP events: impulsive SEP events related to type III radio bursts and gradual SEP events related to type II radio bursts (Reames, 2013). Impulsive SEP events are related to short duration solar flares. Hereafter only the gradual SEP events are discussed because their amplitude is larger than that of the impulsive SEP events.

The shocks ahead of CMEs in the corona and the solar wind are excellent accelerators of charged particles as discussed about supernovae in Chapter 1 (Eqs. 1.11–1.12). Kahler (1978) suggested that CME shocks cause SEP events. In fact, CME speed and SEP intensity are correlated (Gopalswamy et al., 2004), although the scatter from the regression line is large. Gopalswamy et al. (2005) presented a good example to show the relationship between CMEs and SEP events (Fig. 2.9).

The first SEP event was reported by Forbush (1946), which is now called *ground level enhancement* (GLE) event because the GeV protons

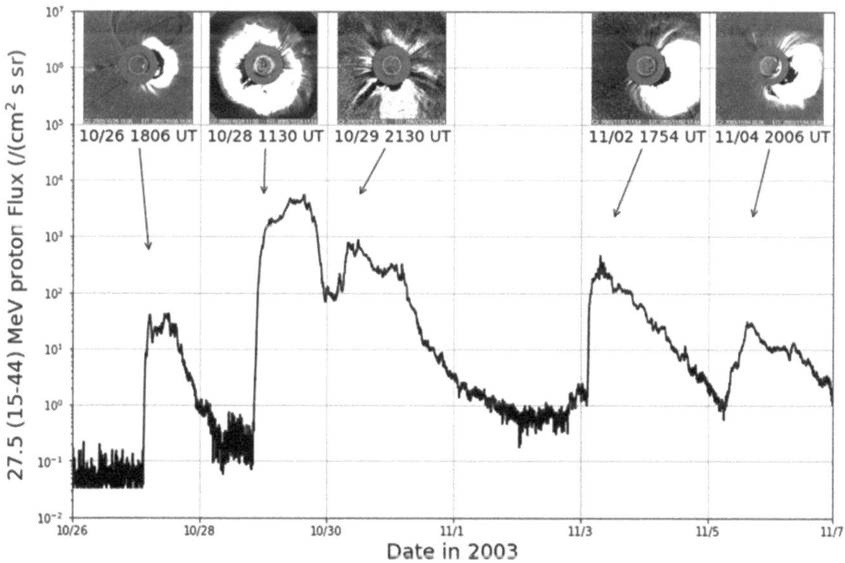

Fig. 2.9 ***Coronal mass ejections (CMEs) causing solar energetic particle (SEP) events in October/November, 2003.*** See also Gopalswamy et al. (2005).

enhance the neutron count rate at ground by creating the air shower. GLE is the most energetic "hard" population of SEP events. SEP events and GLE events tend to occur when CME speed exceeds 1500 and 2000 km/s, respectively (Gopalswamy, 2016). Various energy spectra have been observed during SEP events. For example, the high-flux SEP event which causes of the radiation dose at flight altitude is not necessarily cause GLE event if it was "soft" without GeV protons (Kataoka et al., 2015b).

The maximum energy of shock-accelerated particles with charge Z can be evaluated by multiplying the motional electric field VB and shock size R as follows:

$$E_{\mathrm{max}} = \frac{3}{20}\frac{1}{\xi} ZeVBR, \tag{2.1}$$

$$E_{\mathrm{max}} = 9\times 10^{9}\left(\frac{\xi}{10}\right)^{-1}\left(\frac{V}{4\times 10^{3}\,\mathrm{km\ s^{-1}}}\right)\left(\frac{B}{10^{-5}\ \mathrm{T}}\right)\left(\frac{R}{0.1\ \mathrm{AU}}\right)\mathrm{eV}, \tag{2.2}$$

where the Bohm limit is $\xi = 1$, but it is typically assumed to be 10. GLE events must be accelerated at shocks in the low-altitude corona in which coronal magnetic field is strong. In fact, the time-of-flight analysis of observed GLE particles has the source region at 2 to 3 solar radii (Reames, 2009).

It is also noteworthy that Forbush (1946) reported another interesting phenomenon that cosmic ray count rate decreased a few days after GLE event, which is called *Forbush decrease* events. The Forbush decrease events are caused by the additional magnetic barrier around the Earth against GCR, due to the arrival of interplanetary shocks and CMEs.

The energy deposition of the air shower gives the ion-pair production rate if divided by the W value (34 eV) of the atmosphere. The energy deposition due to SEP and/or radiation belt electrons are the major cause to increase the nitrogen oxide content in the middle atmosphere. Following the ionization, atmospheric chemical reactions produce odd hydrogen (HOx) such as H, OH, and HO_2, and odd nitrogen (NOx) such as N, NO, and NO_2 (Verronen et al., 2011). Increased HOx and NOx concentration causes ozone depletion in the middle atmosphere through catalytic reaction cycles (Crutzen, 1970; Grenfell et al., 2006), as briefly summarized below.

Free electrons interact with nitrogen molecules via electron impact dissociation. Accordingly, the dominant process after the dissociation of nitrogen molecules becomes:

$$N + O_2 \rightarrow NO + O \tag{2.3}$$

Such a NOx enhancement destructs the ozone through the following catalystic reactions:

$$NO + O_3 \rightarrow NO_2 + O_2 \tag{2.4}$$

$$NO_2 + O \rightarrow NO + O_2 \tag{2.5}$$

Thus, the net result of Eqs. (2.4) and (2.5) becomes:

$$O_3 + O \rightarrow 2O_2 \tag{2.6}$$

The simulation results of the ozone destruction by strong SEP events are shown in Fig. 2.10 (Jackman et al., 2005). During the Halloween event, the ozone depletion regions appeared over high geomagnetic and

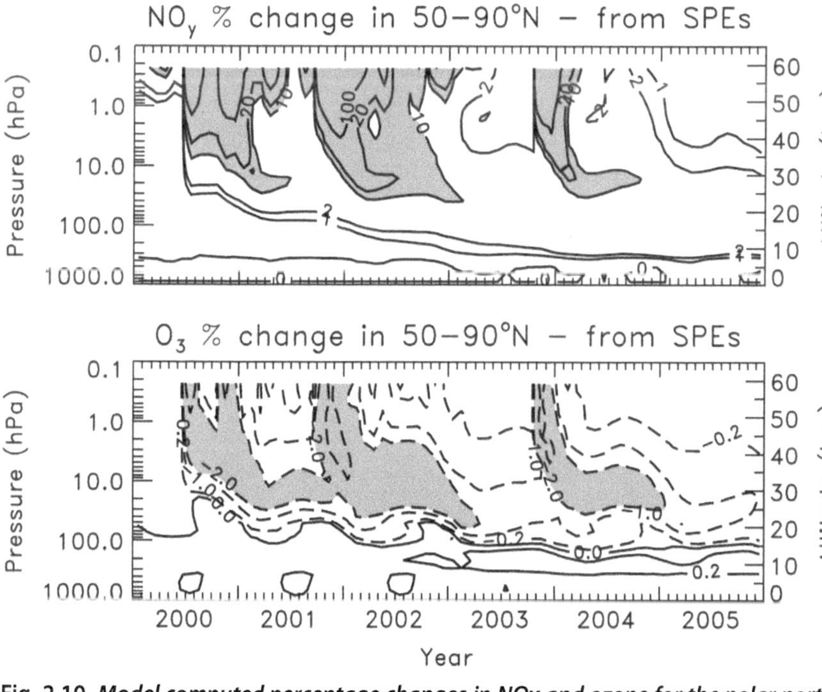

Fig. 2.10 ***Model computed percentage changes in NOx and ozone for the polar northern hemisphere resulting from SEP events (Jackman et al., 2005).***

high geographic latitudes because the polar vortex confined the ozone-depleted atmosphere (Rohen et al., 2005). Bornebusch et al. (2010) investigated the SEP flux during the Halloween event to discuss the inhomogeneities and asymmetries in the precipitation flux in the polar cap region. The asymmetry of the SEP flux in northern and southern polar cap was generally discussed in terms of magnetic topology (Van Allen et al., 1971).

The SEP storm is categorized according to S-scale at NOAA/SWPC (Table 2.2). Since the discovery of GLE events in 1942, a total of 72 GLE events have been registered till 2020 (Table 2.3). Since the monitoring observation by GOES spacecraft in 1975, a few hundreds of large SEP events have also been registered. The largest GLE was observed on February 23, 1956, where the neutron counts increased by 9000% from the pre-event background level (Duggal, 1979; Usoskin et al., 2020). The second largest GLE was observed on January 20, 2005, where the count rate increased by 4200% (Bieber et al., 2013). The time-integrated proton flux for each GLE event is known as *fluence*. The fluence values and energy spectra for 72 GLE events are summarized by Cliver et al. (2020b). The variety of SEP and GLE events and their consequences have been experienced and the knowledge was accumulated for several decades. However, we do not know the possible consequences of extreme SEP events, such as one in 100- or 1000-year events.

Cosmogenic radionuclides provide important hints to understand the possible magnitude and occurrence probability of such extreme SEP events. In the air shower caused by SEP or GCR, the ^{14}C isotope is created when a neutron of nitrogen atom is replaced by a proton. The radioactive half-life of ^{14}C is 5730 years. The ^{14}C accumulates in tree rings every year through the atmospheric carbon cycle and photosynthesis; hence, we can reconstruct the past production rate via tree rings. Super SEP events have been discussed based on the large spike in the ^{14}C proxy data of tree rings in 774–775 AD (Miyake et al., 2012).

Table 2.2 S-scale of NOAA/SWPC to categorize the impact of solar energetic particle (SEP) events.

Scale	S1	S2	S3	S4	S5
Description	Minor	Moderate	Strong	Severe	Extreme
Physical measure of 10 MeV PFU	10	10^2	10^3	10^4	10^5
Average frequency per 11 years	50	25	10	3	Fewer than 1

Table 2.3 List of ground level enhancement (GLE) events from 1942 to 2020. The format of date is year/month/day.

#	Date	#	Date	#	Date
1	1942/2/28	25	1972/8/7	49	1990/5/26
2	1942/3/7	26	1973/4/29	50	1990/5/28
3	1946/7/25	27	1976/4/30	51	1991/6/11
4	1949/11/19	28	1977/9/19	52	1991/6/15
5	1956/2/23	29	1977/9/24	53	1992/6/25
6	1956/8/30	30	1977/11/22	54	1992/11/2
7	1959/7/17	31	1978/5/7	55	1997/11/6
8	1960/5/4	32	1978/9/23	56	1998/5/2
9	1960/9/3	33	1979/8/21	57	1998/5/6
10	1960/11/12	34	1981/4/10	58	1998/8/24
11	1960/11/15	35	1981/5/10	59	2000/7/14
12	1960/11/20	36	1981/10/12	60	2001/4/15
13	1961/7/18	37	1982/11/26	61	2001/4/18
14	1961/7/20	38	1982/12/8	62	2001/11/4
15	1966/7/6	39	1984/2/16	63	2001/12/26
16	1967/1/28	40	1989/7/25	64	2002/8/24
17	1967/1/28	41	1989/8/16	65	2003/10/28
18	1968/9/29	42	1989/9/29	66	2003/10/29
19	1968/11/18	43	1989/10/19	67	2003/11/2
20	1969/2/25	44	1989/10/22	68	2005/1/17
21	1969/3/30	45	1989/10/24	69	2005/1/20
22	1971/1/24	46	1989/11/15	70	2006/12/13
23	1971/9/1	47	1990/5/21	71	2012/5/17
24	1972/8/4	48	1990/5/24	72	2017/9/10

Usoskin and Kovalstov (2021) summarized the recent efforts that the super SEP events identified in tree rings likely represent the low-probability tail of the continuous occurrence distribution of SEP events.

The spallation of a neutron from an oxygen atom creates another useful radionuclide ^{10}Be in the air shower. The ^{10}Be isotope has a ~1.4 million year half-life and is accumulated in ice cores via aerosol fallouts. The ^{10}Be content represents the fluence of ~200 MeV protons, while the ^{14}C content represents the fluence of ~1 GeV protons. There is another radionuclide, ^{36}Cl, which is sensitive to ~30 MeV protons. The analysis of multiple proxy data suggests that the spike event in 774–775 AD can be the historically largest SEP event, including the possibility that multiple SEP events occurred that year (Usoskin et al., 2013). From the solar cycle modulation found in ^{10}Be, the 774–775 AD event likely occurred around the solar maximum (Sukhodolov et al., 2017). Also, it is notable that nitrate deposition signal is indistinguishable from the seasonal cycle even for the largest SEP events (Sukhodolov et al., 2017), although the nitrate composition in ice cores have been used as proxy of large SEP events (Cliver and Svalgaard, 2004).

The proxy data are only sensitive to the largest SEP events. The largest event occurred in 774–775 AD, which can be ~140 times larger than the January 20, 2005 SEP/GLE event (Mekhald et a., 2015), while the second largest and third largest events occurred in 660 BC and 993–994 AD, which are ~100 times and ~50 times larger than the January 2005 event, respectively. Note also that the largest (soft) SEP event was observed on August 4, 1972, while the largest (hard SEP or) GLE event was observed on February 23, 1956 (Fig. 2.11).

What is the worst case of the largest SEP event? As a simple answer, SEP events with an order of magnitude larger than the 774–775 AD event are unlikely, as concluded from the 10-year resolution tree ring ^{14}C data during the last 11 thousand years of Holocene. Cliver et al. (2020a) have discussed that the 774–775 AD event can be converted to X200 or 10^{33} erg solar flares. Note also that so far we experienced up to 10^{32} erg event on November 4, 2003. Assuming such a simple conversion of the SEP proxy data to X-ray intensity, the occurrence of super flares beyond 10^{34} erg is therefore unlikely in our Sun, although Shibata et al. (2013) have discussed that there is a possibility that 10^{34} erg super flare occur once in 800 years on our present Sun. Sukhodolov et al. (2017) discussed that the

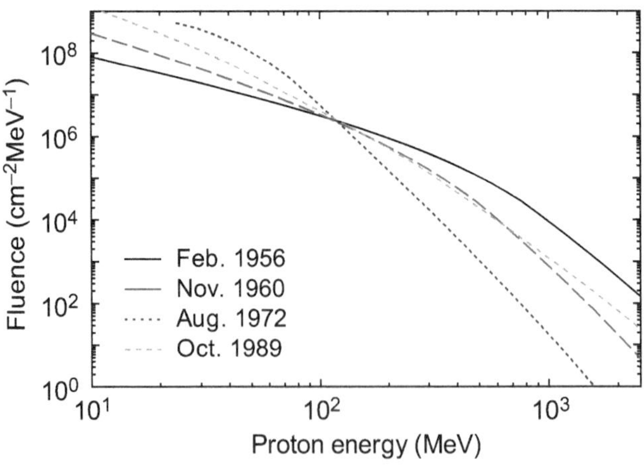

Fig. 2.11 ***Energy spectra of the largest SEP events.*** See also Tylka and Dietrich (2009).

774–775 AD SEP event can decrease the stratospheric ozone for more than one year and thus can even modulate the surface weather.

2.4 Geomagnetic disturbances (GMD)

Auroras are the flare of the Earth, and the explosive energy release in the magnetosphere is called *substorm*. The energy release is two-fold. One is the Earth version of CMEs, called *plasmoid*, which directs away from the Earth. Another is the bursty plasma injection toward the Earth to cause the *auroral substorm*, which is a core element to understand the various phenomena in the ionosphere and magnetosphere. Akasofu (1964) summarized that the basic sequence of auroral substorm by analyzing all-sky images of auroras captured worldwide (Fig. 2.12). Similar auroral oval explosions occur in both northern and southern polar regions, and such a macroscopic view helps to understand that the auroral substorm is the Earth version of solar flare.

Once the auroral substorm occurs, various auroral phenomena are manifested in the ionosphere. The beautiful fine-scale dynamic auroras

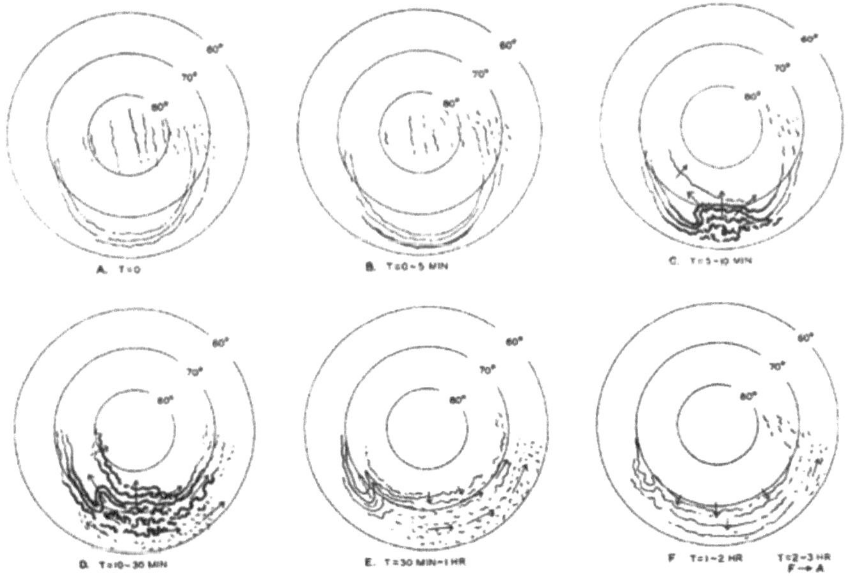

Fig. 2.12 ***Schematic diagram of auroral substorm in the northern polar region at magnetic latitude above 60 degrees (Akasofu, 1964).*** After an initial brightening at the equatorward edge of auroral oval (B), auroral breakup, and westward traveling surge occurred in the expansion phase (C), then diffuse and pulsating auroras appear postmidnight local time (D–F). To complete the substorm cycle, the growth phase exists before (A).

appear during the substorm breakup (Kataoka et al., 2021). The human eyes cannot fully capture the time variations of auroras. For example, *flickering auroras* are known as the extremely fast intensity variation appeared in a part of bright discrete auroral arcs. Thanks to the recent advance of sensitive high-speed cameras, ~80 Hz flickering has been captured (Fukuda et al., 2017), although McHarg et al. (1998) reported the possible existence of ~180 Hz flickering. Diffuse and pulsating aurora then appears after the breakup. The on-off time sequence of a few seconds time scale, and the 3 Hz intensity modulation of pulsating auroras are caused by chorus wave intensity variations (Nishimura et al., 2020; Hosokawa et al., 2020). The fastest modulation of pulsating aurora is ~60 Hz (Kataoka et al., 2012b).

For a few hours after the auroral breakup, auroral electron beams ionize the polar atmosphere to enhance the electric conductivity. The horizontal ionospheric current is therefore largely enhanced along the expanding auroral oval. This is called *auroral ejectrojet*. The electric current system of auroral electrojet carries more than million Ampere, and the Joule heating rate can be more than Tera Watt (Shiokawa et al., 2007). By Ampere's law, the auroral electrojet causes the large-amplitude *geomagnetic disturbance* (GMD) at ground, and the deviation in the horizontal component often exceeds 10^3 nT.

The substorm is driven by the large energy flow from the solar wind into the magnetosphere. The energy flow is maximized when the *interplanetary magnetic field* (IMF) is just opposite to the geomagnetic field (Fairfield and Cahill, 1966). Since the geomagnetic field points northward, the southward-directing IMF matters, which is now called SBZ. It has been intuitively interpreted that SBZ and magnetic reconnection may drive the enhanced plasma convection in the "open" magnetosphere (Dungey, 1961).

The substorm energy dissipation in the atmosphere is typically 0.1 to 0.5 TW. Those values can be estimated by integrating the joule heating rate over ploar regions. The energy accepted from the solar wind increases with the increase in magnetosphere size. The additional energy into the atmosphere via the Joule heat can contribute to flow out a significant part of the upper atmosphere (Nilsson, 2011), although the global magnetic field is also important to keep the atmosphere against the direct interaction by the solar wind. From this electromagnetic contribution to the atmospheric loss, whether having a strong geomagnetic field works positively for keeping the atmosphere is uncertain.

In general, GMD have a wide range of temporal scale, from seconds to days. One of the most rapid and large amplitude variations is caused by a sudden magnetosphere compression due to the arrival of the interplanetary shock. This is called SC, which is identified as the rapidly increased northward component of the geomagnetic field at ground (Araki, 1994). These phenomena are also called *sudden impulse*, and particularly called *storm sudden commencement* (SSC) if the magnetic storm follows the SC. The typical temporal scale of SC events is seconds to minutes. An anomalous SC occurred on March 24, 1991, where the amplitude and rise time were 202 nT and 28 seconds, respectively (Araki et al., 1997). Larger SC with 220 nT amplitude was observed on 13, November 1960, while the largest SC with 273 nT amplitude was observed on March 24, 1940 (Araki, 2014).

A part of GMD is categorized by the frequencies and waveforms into continuous pulsations (Pc) and irregular pulsations (Pi), both of which have been widely used to diagnose the magnetosphere (Saito, 1969). These *geomagnetic pulsations* are often called ultra-low frequency (ULF) waves (Table 2.4). Among the ULF waves, Pc5 and Pi2 are frequently used in space weather studies. Pc5 is frequently discussed as the possible driver to cause the radial diffusion of the radiation belt electrons, while Pi2 is used as the substorm onset marker.

The slowest variation, progressing in hourly time scale, is called *magnetic storms*, which is identified by the decreased northward component of the geomagnetic field at ground. Prolonged SBZ for several hours can accumulate the energetic particles in the inner magnetosphere. This is identical to *ring current* evolution, which is called the *main phase* of a magnetic storm. Simultaneously, the accumulated energetic particles suffer from various loss processes. When the source process such as a series of substorms starts to cease, the ring current gradually disappears, which is called the *recovery phase*.

SBZ is therefore the most important driver for causing not only substorms but also magnetic storms, although the progressing temporal scales are different. In other words, the arrival of a fast CME is not sufficient to cause major geomagnetic disturbances, and SBZ is necessary to drive both substorms and magnetic storms. The importance of SBZ also appears in the seasonal variations in GMD intensity, i.e. large-amplitude events tend to occur in spring and fall. A large portion of the seasonal variation can be

Table 2.4 Geomagnetic pulsations categorized as ultra-low frequency (ULF) waves.

ULF class	Pc 1	Pc 2	Pc 3	Pc 4	Pc 5	Pi 1	Pi 2
Periods (s)	0.2–5	5–10	10–45	45–150	150–600	1–40	40–150

understood by the geometrical setting of the inclined geomagnetic field against the Parker spiral of the solar wind (Russell and McPherron, 1973). The Russell–McPherron effect is the main reason why the coordinates of Geocentric Solar Magnetospheric System (GSM) are used for evaluating SBZ, rather than Geocentric Solar Ecliptic (GSE).

Magnetic storms are ranked by the Dst index since 1957 (Sugiura, 1964; Sugiura and Kamei, 1991). The Dst index is derived from 1-hour averaged north-south magnetic field variation of four low-latitude ground-based magnetometers. During magnetic storms, the ring current development in the inner magnetosphere decreases the Dst index. The SYM-H index, 5 minutes version of the Dst index, is often used to identify SC events. The high-latitude substorm activity is ranked according to the auroral electrojet (AE) index. Kp index or NOAA G-scale is also often used to categorize the GMD activity (Table 2.5). The 3-hour GMD activity is ranked by 28 levels (0, 0+, 1−, ..., 9−, 9) as K-index at each sub-auroral magnetometers (50°–60° magnetic latitude), and the weighted average over the world is the Kp index (Mayaud, 1980). Matzka et al. (2021) have reviewed the property of Kp index.

These different geomagnetic activity indices show the maximum values at different timings, as they represent different current systems. In short, Dst represents the ring current, Kp roughly represents the sub-auroral current system, and AE represent the high-latitude ionospheric currents. Fig. 2.13 provides an interesting example that the strong AE activity was associated with the shock downstream SBZ (sheath SBZ), while the Dst index developed afterward when prolonged SBZ (cloud SBZ) started. The Kp index responded to both sheath and cloud regions.

Outstanding magnetic storms (Kp = 9) since 1932 are listed in Table 2.6. Note that Kp = 9 was defined to include even the exceptionally large amplitude event at 0600 to 0900 UT on April 16, 1938 (Sergeyeva et al., 2021), exceeding 2500 nT in the auroral zone and about 300 nT at low latitudes. Magnetometer at Lerwick observed the extreme horizontal field deviation of 2980 nT during the extreme storm on April 16, 1938 (The

Table 2.5 G-scale of NOAA/SWPC to categorize the impact of geomagnetic disturbances (GMD), associated with both substorms and magnetic storms.

Scale	G1	G2	G3	G4	G5
Description	Minor	Moderate	Strong	Severe	Extreme
Physical measure (Kp)	5	6	7	8	9
Average frequency per 11 years	1700	600	200	100	4

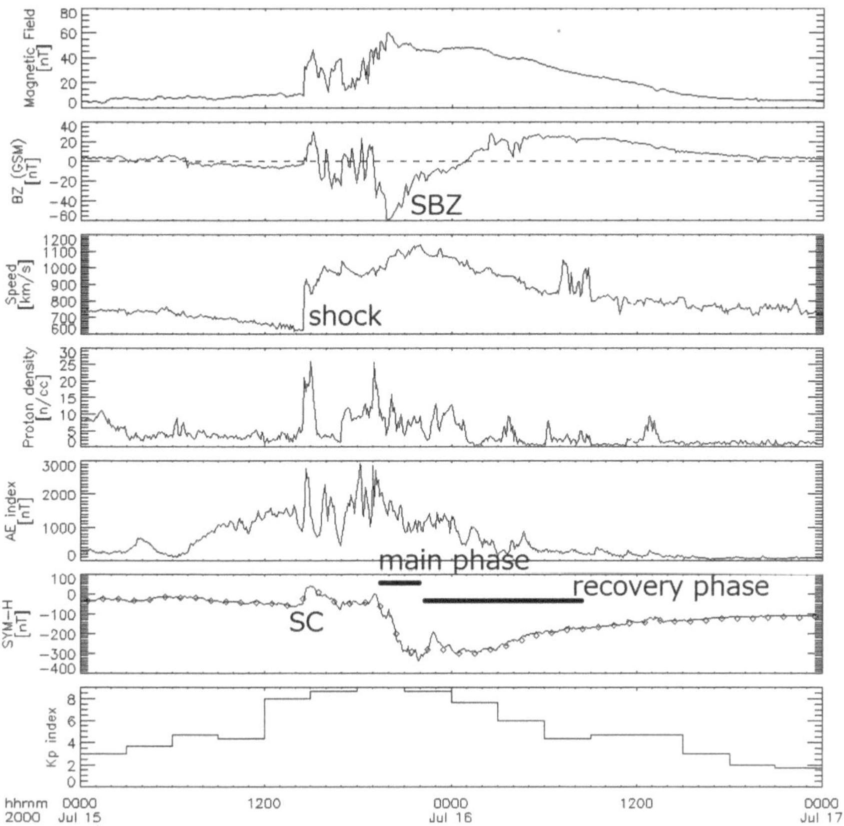

Fig. 2.13 ***Example data of the Bastille Day event to show that southward-directing magnetic field (SBZ) in the solar wind is the important driver for both substorms and storms.***

Magnetic Storm of April 16. Nature 141, 721-722, 1938). Before 1932, outstanding magnetic storm events also occurred on February 4, 1872, February 14, 1892, September 25, 1909 (reconstructed Dst = −595 nT, Love et al., 2019a), and May 14–15, 1921 (reconstructed Dst = −907 nT, Love et al., 2019b).

Two different drivers cause magnetic storms by holding strong SBZ. Recurrent storms are often driven by *corotating interaction region* (CIR), which are called CIR-driven storms. The recurrent storms are also called *gradual storms* as we do not often see SC. The intensity of CIR storms is relatively small and rarely exceed −Dst ~200 nT (Richardson et al., 2006). The large amplitude magnetic storms are CME-driven. They are also called SSC (storm sudden commencement) storms because the CME-driven

Table 2.6 Outstanding magnetic storms from 1932.

Extreme GMD (Kp = 9) events (1932/1–2020/12)				
Year	**Month**	**Day**	**Hour**	**Dst peak (nT)**
1938	4	16	6	
1940	3	24	15,18	
1941	3	1	15	
1941	7	5	12	
1941	9	19	3	
1946	7	27	6	
1946	9	22	12	
1957	9	4	15	−324
1958	2	11	0	−426
1958	7	8	15	−330
1959	7	15	15	−429
1960	4	30	15	−325
1960	10	7	0	−287
1960	11	13	6,9	−339
1967	5	25	21,24	−387
1970	3	8	18	−284
1972	8	4	21	−118
1982	7	13	21	−325
1982	7	14	0	−325
1986	2	8	18	−307
1989	3	13	21,24	−589
2000	7	15	18	−300
2003	10	29	6	−353
2003	10	30	18,21	−383

Dst peak values are also noted after 1957.

storms are often preceded by SC. The interaction among multiple CMEs and CIR leads to further compression and therefore strong SBZ, causing super magnetic storms (Liu et al., 2014; Kataoka et al., 2015a).

When CMEs surround the magnetosphere of the Earth, the Alfvén Mach number of the solar wind significantly reduces because of the strong magnetic field (e.g., Kataoka et al., 2005b). The Alfvén Mach number ranges from 7 to 11 in a typical high-speed solar wind, while it decreases to less than the half during CMEs. In this special situation of low Mach number, the bow shock was observed far away from the magnetopause (Fairfield et al., 2001). The low-Mach solar wind–magnetosphere interaction can be different from the usual high-Mach state, because of the weak bow shock and resultant low-beta flow in the magnetosheath. Global MHD simulations predict that due to the relatively strong magnetic force, the cross-section of the magnetopause shape and the surrounding magnetosheath flow becomes asymmetric (Lavraud and Borovsky, 2008). Lavraud et al. (2007) have also reported in-situ observation data that the magnetosheath flow was beyond 1000 km/s against the upstream solar wind speed of 650 km/s, when the Mach number was between 2.0 and 4.4. Other low-Mach examples include inner magnetospheric response (Lugaz et al., 2016).

During CME arrival, the magnetospheric response becomes nonlinear to solar wind energy input. The most popular parameter to measure such nonlinearity is the cross polar cap potential (CPCP), which is defined as the maximum dawn-to-dusk voltage drop applied to the northern or southern polar ionosphere. The CPCP is usually proportional to the solar wind electric field, Esw = Vsw × SBZ, but it is smaller than the expected strength when Esw increases significantly (Shepherd, 2007). Recent MHD simulation results indicate that CPCP saturation is also dependent on Alfvén Mach number (Lakka et al., 2018).

In the extreme situation of sub-Alfvenic ($M_A < 1$) solar wind, the magnetosphere shows an interesting shape called Alfvén wings (Chane et al., 2012), instead of the usual comet tail shape. The Alfvén wing formation is a proposed mechanism to explain CPCP saturation (Ridley, 2007; Kivelson and Ridley, 2008). Other proposed mechanisms include decreasing the subsolar magnetopause reconnection rate (Hill et al., 1976) and the limitation from the Region 1 current system (Siscoe et al., 2002, 2004).

To understand the extreme GMD, it is important to look at the local variations, beyond the indices. As the worst case, a low-latitude magnetometer recorded ~1600 nT spike (Fig. 2.14) during the Carrington magnetic storm on September 2, 1859 (Tsurutani et al. 2003). However, Siscoe et al. (2006) estimated the 1-hour averaged value of −850 nT as the reconstructed Dst index for this data. Modeling work was also performed by Li et al. (2006). It is also notable that Akasofu and Kamide (2005) claimed that there are several reasons to believe that the 1600 nT amplitude spike in Fig. 2.14 is unrealistic to be a Dst value. Cid et al. (2015) argued that field-aligned currents in middle latitudes may contribute to the spikes. In any case, those amplitudes are well below the theoretical upper limit of −Dst, which is estimated to be 2500 nT (Vasyliunas, 2011, 2013). Even though such large local peaks do not result in a dramatic global response as seen in Dst, local disturbances can have nevertheless significant space weather hazards, and it is important to understand the possible mechanisms to cause the large peaks.

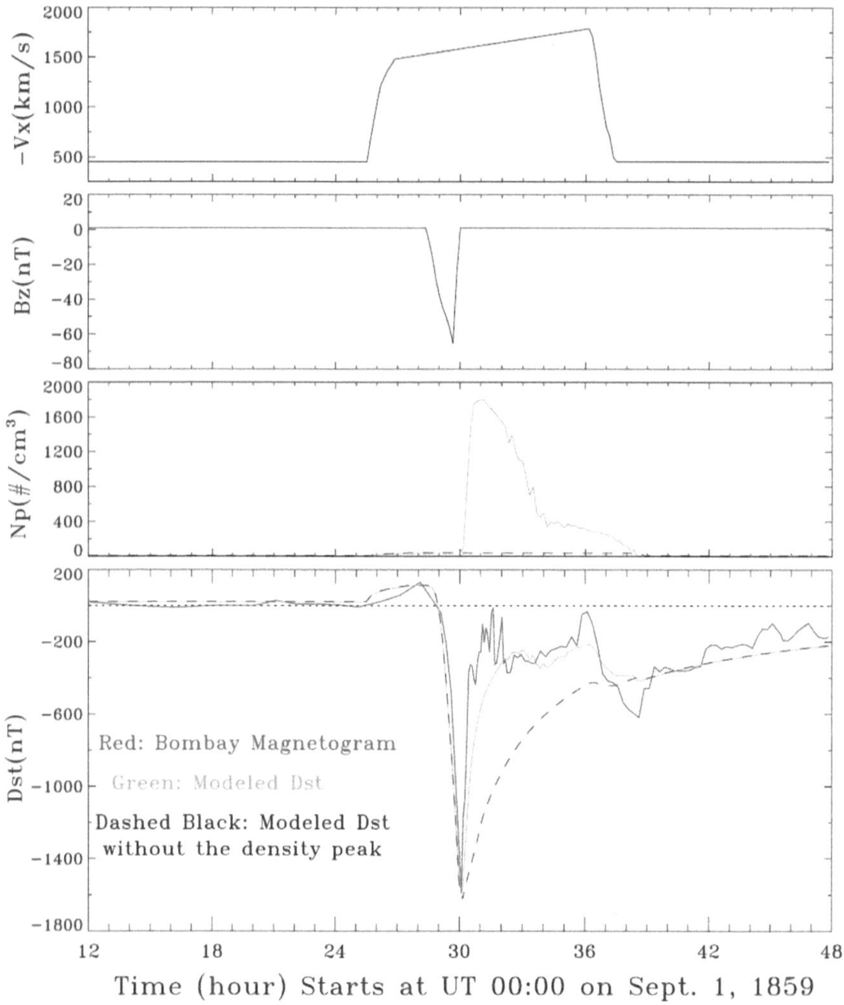

Fig. 2.14 ***Carrington magnetic storm observed at Colaba observatoary in Mumbai in 1859, and assumed solar wind parameters to model the magnetic storm (Li et al., 2006).*** The pioneering work was given by Tsurutani et al. (2003).

From the statistical comparison among several space weather phenomena of magnetic storms, solar flares, and CMEs, Riley (2012) estimated the probability of another Carrington event in the next 10 years to be 12%. Kataoka (2013) applied the same analysis to the 90-year list of magnetic storms and estimated the probability of another Carrington storm in next 10 years to be 4 to 6%.

Predicting the equatorward extent of the auroral oval during extreme magnetic storms is important to prepare a global hazard map for possible

GIC accidents. As an extreme case, Green and Boardsen (2006) argued the global auroral extent around the Carrington storm from August 28 to September 3, 1859, as follows: At the peaks of geomagnetic activity, red auroras appeared at ~25° magnetic latitude on August 28–29, and at ~18° magnetic latitude on September 2–3. Auroral forms of all types and colors were observed below 50° magnetic latitude for ~24 hours on August 28–29 and ~42 hours on September 2–3. Silverman (2008) discussed that the Carrington event is not likely the most extreme case of low-latitude auroral extent because the aurora was observed at low-latitude region of ~10° magnetic latitude, such as Bombay, India, on February 4, 1872 (Chapman, 1953).

It has been discussed that 40° magnetic latitude is a good measure of extremely extended equatorward auroral boundary (Ngwira et al., 2013b; Kataoka and Nakano, 2021). Extending the work by Yokoyama et al. (1998), Kataoka and Nakano (2021) updated the empirical relationship between the magnetic latitude of auroral equatorward boundary and the Dst index, concluding that the efficiency to transform the magnetic energy to the ring current energy can be ~20% for great storms (Fig. 2.15).

The amplitude of magnetic storms, as represented by the Dst index, can be converted as the accumulated energy of ring current particles in the inner magnetosphere. Dessler–Parker–Sckopke equation (Dessler and Parker, 1959; Sckopke, 1966) relates Dst variation with ring-current particle energy as:

$$\frac{D_{st}}{B_0} = \frac{2}{3}\frac{E_R}{E_m}, \tag{2.7}$$

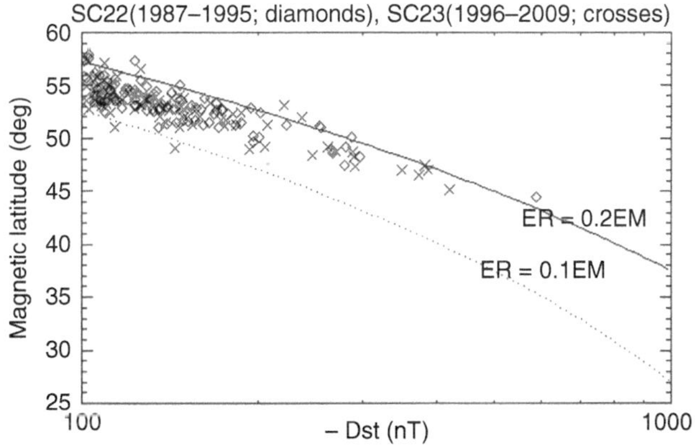

Fig. 2.15 ***Auroral equatorward boundary and peak Dst index during magnetic storms.*** Dotted and solid lines are theoretical curves assuming 10 and 20% energy transform from the ambient magnetic field to the ring current particles, respectively (Kataoka and Nakano, 2021).

where B_0 is the surface magnetic field intensity at the equator, E_R is the sum of the kinetic energy of ring current particles, and E_m is the total energy of the magnetic field external to the Earth, defined by:

$$E_m = \frac{4}{3}\pi R_E^3 \frac{B_0^2}{\mu_0} = 8 \times 10^{17}\,\mathrm{J} \tag{2.8}$$

It is possible that we will have extreme events even larger than the Carrington storm in future. For example, from the detailed analysis of an old auroral painting (Fig. 2.16), Kataoka and Iwahashi (2017) estimated that the amplitude of a historical magnetic storm on 17 September 1770 can be comparable to or even larger than that of the Carrington storm. They assumed a relationship between the equatorward auroral extent and magnetic storm amplitude (Fig. 2.15). The similar fan-shaped aurora appearance was independently observed (Fig. 2.17). Such a fan shape can be the most dramatic auroral form during super magnetic storms in the mid-latitude region, based on the very first auroral photograph data

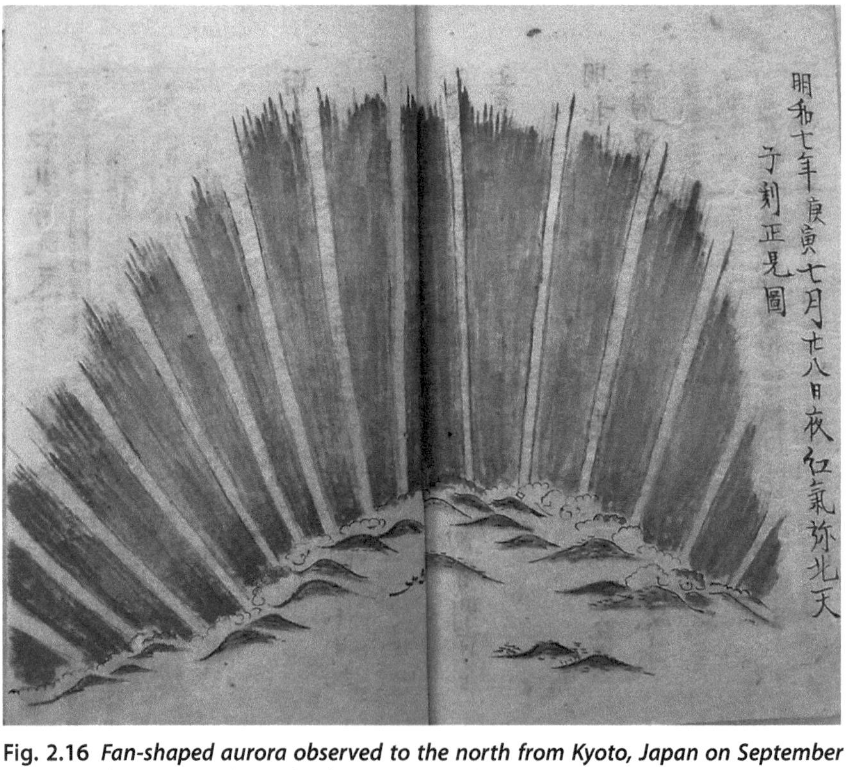

Fig. 2.16 ***Fan-shaped aurora observed to the north from Kyoto, Japan on September 17, 1770.*** The magnetic latitude of Kyoto was 24 degrees (Kataoka and Iwahashi, 2017).

Fig. 2.17 ***Painting by Trouvelot entitled "Aurora Borealis.*** As observed March 1, 1872, at 9h. 25m. P.M." *(Plate IV from The Trouvelot Astronomical Drawings 1881).*

in Japan during the great magnetic storm event on February 11, 1958 (Kataoka et al., 2019; Kataoka and Kazama, 2019). Moreover, a variety of auroras has also been reported in East Asia, as summarized by Willis et al. (2007).

As the latest example, the amplitude of a magnetic storm in May 1921 was estimated to be comparable to that of the Carrington event (Hapgood 2019; Love et al. 2019b), suggesting that the Carrington-class events may be more frequent than previously expected. During this May 1921 event, aurora was observed overhead at 40° geomagnetic latitude, and an aurora was observed from Apia, Samoa, in the southern hemisphere within 13° of the geomagnetic equator (Silverman and Cliver, 2001).

The snapshot image of extended auroral oval over U.S. and the time variation of March 1989 storm are shown by Boteler (2019). A satellite image at 1826 UT on March 13, 1989 showed the larger-than-Antarctica auroral oval. At that time, auroral sightings were also reported across Austraria and Tasmania (Allen et al., 1989).

In Japan, the witness records of possible red aurora have been documented over 1000 years, which are useful to understand the nature of extreme storms. The oldest red aurora event was recorded in *Nihon-shoki* as the earliest astronomical event in Japan as follows (Kataoka et al., 2020b); "On December 30 in 620, a red sign appeared in heaven. The length was more than one jo (10 degrees). The shape was similar to a pheasant tail (Suiko-Tennou, 28)."

The oldest successive red-sign events, that is, multiple occurrence of red auroras for a few days, can be found in *Meigetsuki* (Kataoka et al, 2017) as follows; "On 1204/02/21, it was sunny. … After sunset, red sign appeared in the direction of north and north-east. The lower part of red sign was shaped like a rising moon and colored white and bright. Its stripes extended faraway and were like smokes in fires. There were four or five of white parts and three or four of red sign appeared. Is it neither cloud nor stands within clouds? Its light did not get darken and red light is mixed in the white light. It is nothing but a mystery. It is also very dreadful. On 1204/02/23, it was sunny and quite windy. … In the time when to put a fire on the lamp, red vapor appeared in the north and north-east. It was like a distant mountain burning. It was very dreadful."

The successive red-sign events implies that the active Sun erupted multiple CMEs even 800 years ago, similarly to the situation of Halloween storms in 2003. On the same dates, on 21-23 February 1204, it was independently documented in the national treasure *Omurosoushouki* that people postponed the pilgrimage, regarding the rare red-sign events as a bad omen. It is an interesting evidence of the earliest action of Japanese people against the extreme space weather. It is also notable that these events in 620 and 1204 occurred when the magnetic latitude of Japan was relatively high in the last 3000 years (Kataoka and Nakano, 2021).

2.5 Radiation belts

Radiation belts consist of the relativistic electrons and protons trapped in the geomagnetic field. The outer belt is dominated by relativistic electrons of an order of MeV, while the inner belt is dominated by relativistic protons of an order of 10 MeV (Fig. 2.18). The inner belt was discovered by the Explorer 1 of the USA (Van Allen and Frank, 1959), while the outer belt was discovered by the Sputnik 2 of the USSR (Vernov and Chudakov, 1960). The historical story of the discovery of radiation belts in the USA and USSR during the Cold War was briefly summarized by Baker and Panasyuk (2017). The radiation belts were recognized as one of the most outstanding findings

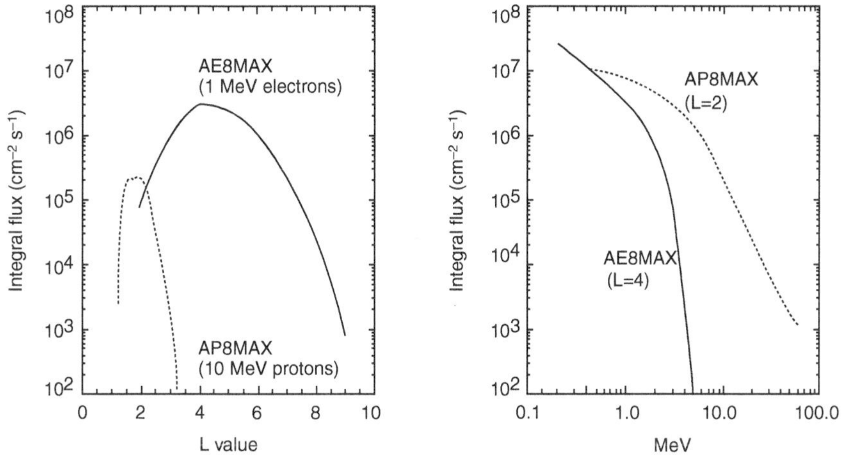

Fig. 2.18 ***Spatial distributions and the energy spectra of the inner radiation belt protons and outer radiation belt electrons at equator during the solar maximum, using AP8 and AE8 models, respectively.***

of the International Geophysical Year (IGY) of 1957–1958. During the IGY, many fundamental scientific observations were conducted worldwide, including the beginning of Japanese Syowa Station in Antarctica.

After the discovery of radiation belts in the beginning of space age, dynamic variations of the outer belt electrons have been intensively studied for more than 60 years. It is evident that high-speed solar wind plays an essential role to enhance the electron flux in the outer belt (Paulikas and Blake, 1979). The good correlation between the solar wind speed and the outer belt electron flux is shown in Fig. 2.19. During the declining phase of the solar cycle, especially in 1994 and 2004, the solar wind speed was high, and the outer belt electron flux was also high due to large coronal holes. During the deep solar minimum of 2009, the solar wind speed was persistently slow, and the MeV electron flux at GEO almost disappeared (Kataoka and Miyoshi, 2010).

Another evident pattern is that the outer belt disappears during storm main phase and is enhanced again during the recovery phase (Fig. 2.20). Both the loss process and source process work simultaneously to make the situation of the outer belt complex.

For example, the flux enhancement of outer belt electrons does not correlate with magnetic storm intensity (Fig. 2.20). Miyoshi and Kataoka (2005) and Kataoka and Miyoshi (2006) showed that the outer belt flux enhancements depend more on the large-scale solar wind structures, CIRs and CMEs. This dependence is seen over solar cycles (Miyoshi and Kataoka,

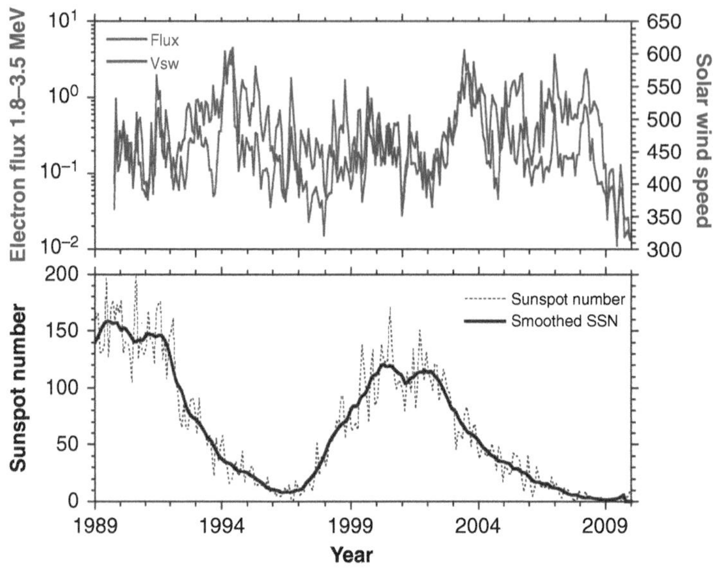

Fig. 2.19 ***Correlation between the electron flux at LANL at GEO and solar wind speed with the solar cycle variation (Reeves et al. 2011).***

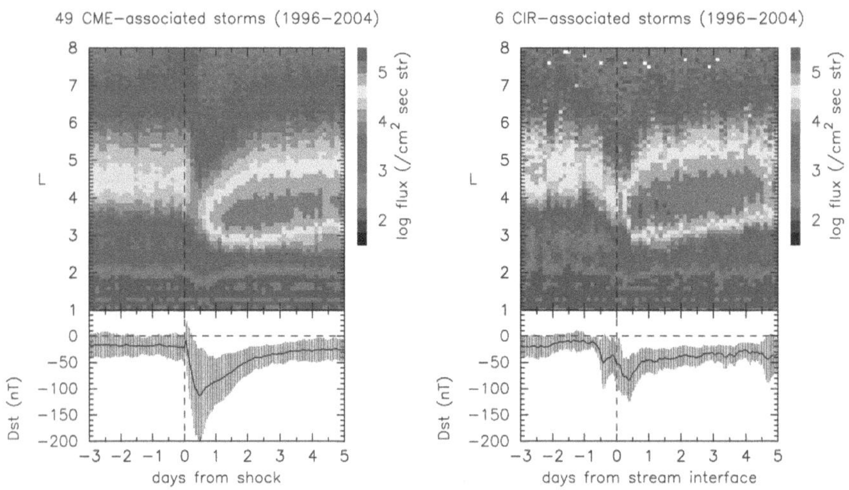

Fig. 2.20 ***Averaged variation of outer radiation belt electron flux against L value during magnetic storms.*** Left and right panels indicate the magnetic storms caused by coronal mass ejections (CMEs) and corotating interaction regions (CIRs), respectively. The zero epochs are interplanetary shocks and stream interfaces, respectively (Kataoka and Miyoshi, 2006).

2011). Particularly strong depletions of outer belt electron flux occur during the passage of CME sheath (Kilpua et al., 2015b).

Miyoshi et al. (2013) showed that the SBZ in the high-speed solar wind affects the key parameters of electron acceleration by chorus waves. These parameters include source electron population, thermal plasma density, and chorus wave activity itself. They showed that prolonged substorm activity associated with the SBZ is essential for the large electron flux enhancement. Jaynes et al. (2015) also pointed out the importance of prolonged substorm activity in gradual acceleration of MeV electrons, by providing both source electron population that excites chorus waves as well as seed electrons that are accelerated by the generated chorus waves. The substorm effects therefore roughly explain that large flux enhancements in the recovery phase of CIR-storms rather than CME-storms (Miyoshi and Kataoka, 2005; Kataoka, and Miyoshi, 2006), because the prolonged substorm activities occur associated with the high-speed coronal hole streams (Tsurutani et al., 2006). It is also notable that fast streams typically lack the condition favoring the loss processes such as magnetopause shadowing (Kilpua et al., 2015b).

The energy density or pressure of radiation belt particles is smaller than the magnetic pressure of the inner magnetosphere, and the approximation of test particles in the ambient magnetic field is appropriate. In the adiabatic condition, the magnetic moment μ of a particle is constant:

$$\mu = \frac{mv_{\perp}^{2}}{2B} \tag{2.9}$$

The quantity μ is also called *first adiabatic invariant*. Adiabatically increased B increases $v_{\perp}^{2}$. Considering also the conservation of kinetic energy in a dipole field, $v_{\parallel}^{2}$ decreases when B increases. The location where the parallel velocity $v_{\parallel}^{2}$ changes the sign is called *mirror point*. The second adiabatic invariant J is associated with the cyclical bounce motion between the mirror points. The third adiabatic invariant is related to the cyclic magnetic drifts in longitudinal direction (Fig. 2.21). Breaking any of the three invariants defines "nonadiabatic" processes.

It is convenient to express the magnetic drifts in terms of the parallel and perpendicular nonrelativistic kinetic energies:

$$K_{\parallel} = \frac{1}{2} mv_{\parallel}^{2} \tag{2.10}$$

$$K_{\perp} = \frac{1}{2} mv_{\perp}^{2} \tag{2.11}$$

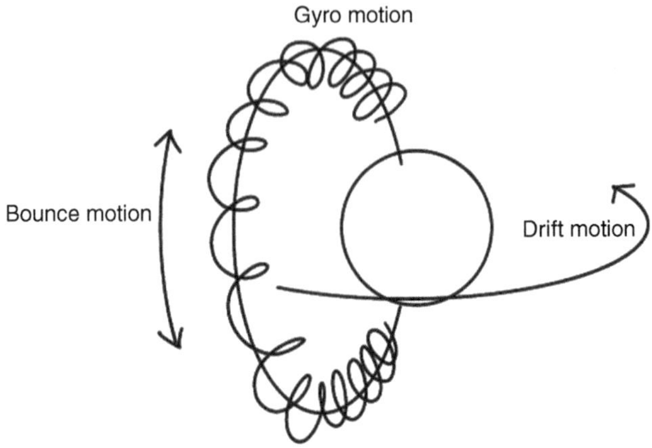

Fig. 2.21 ***Schematic diagram of gyro motion around a magnetic field-line, bounce motion between the northern and southern hemispheres, and drift motion around the Earth.***

The drift velocity of gradient B force then becomes:

$$\mathbf{v}_{grad} = \frac{K_\perp}{qB}\frac{\mathbf{B}\times\nabla B}{B^2} \tag{2.12}$$

and the drift velocity of the magnetic field curvature force becomes:

$$\mathbf{v}_{curv} = \frac{2K_\parallel}{qB}\frac{\mathbf{B}\times\nabla\mathbf{B}}{B^2} \tag{2.13}$$

The general expression of gradient-curvature drift can therefore be written as:

$$\mathbf{v}_{magnetic} = \frac{\left(K_\perp + 2K_\parallel\right)}{qB}\frac{\mathbf{B}\times\nabla B}{B^2} \tag{2.14}$$

Considering these test particle motions, L value or L shell is the most useful way to indicate the location of particles (McIlwain, 1961). Whilst there are several kinds of L values, the L value in the dipole field is calculated by the footprint latitude λ by:

$$L = \cos^{-2}\lambda \tag{2.15}$$

The *South Atlantic Anomaly* (SAA) is the localized area of geomagnetic field depression. The mirror points of energetic particles in this area are therefore lower than that in the other areas, acting as a sink of trapped energetic particles. The existence of such localized modification from the dipole magnetic field makes the particle motion complex. Girgis et al. (2022) performed the test particle simulation (Fig. 2.22). An example of the

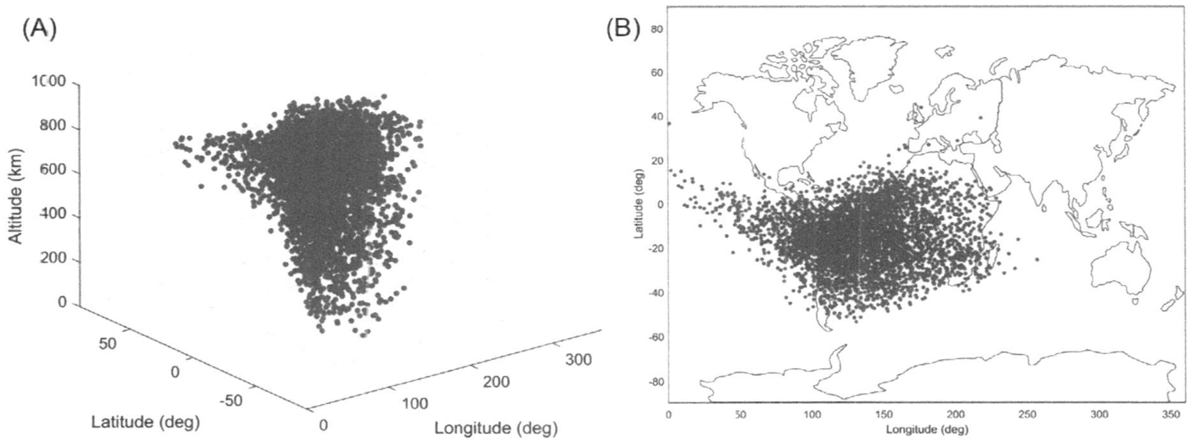

Fig. 2.22 ***Simulated spatial distribution of 70 to 180 MeV protons in geomagnetic field (Girgis et al., 2022).***

dose rate observation at ISS, crossing both SAA and outer radiation belt, was presented by Dachev et al. (2017).

The in-between of the outer belt and the inner belt has been called *slot region*, where the particle flux is supposed to be relatively low. The slot region and the inner belt are rarely disturbed. However, in an extreme example, Combined Release Radiation Effects Satellite (CRRES) once observed a dynamic intrusion of a large amount of MeV electrons and tens of MeV protons into the inner belt on March 24, 1991, to form the double-peaked inner belt (Mullen et al., 1991). The extreme compression of the magnetosphere, that is, the extreme SC (Araki et al., 1997), and resultant variation of the convection electric field explain the newly injected particles in a localized region, and the azimuth drift of the new particle packets called "drift echoes" (Fig. 2.23) explains the

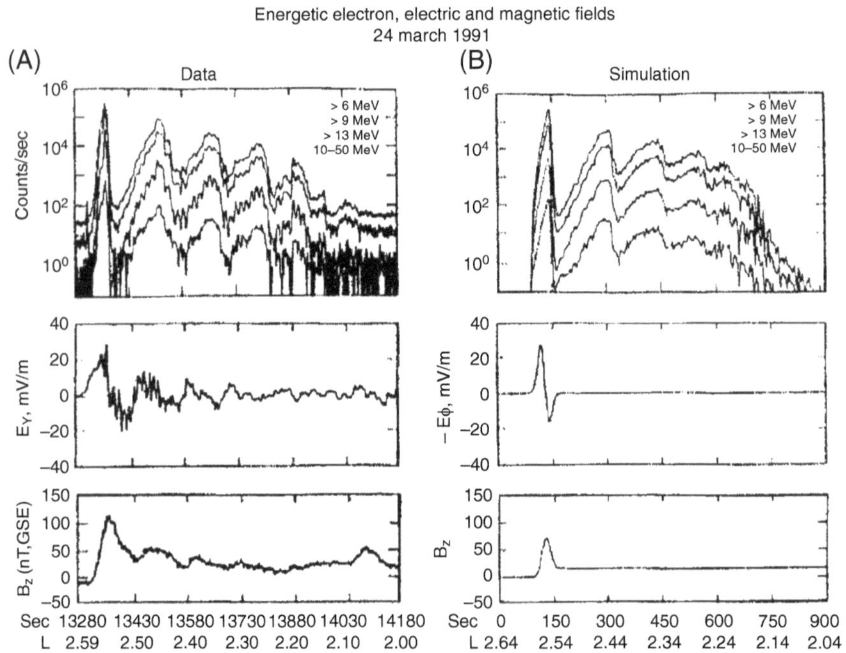

Fig. 2.23 (A) Data from the Combined Release Radiation Effects Satellite (CRRES) satellite during super sudden commencement on March 24, 1991. Top panel shows 6–50 MeV electron flux for 900 seconds time interval. Middle and bottom panels show the measured electric field in a corotational frame and the magnetic field with a model magnetic field subtracted. (B) Simulated results in the same format as (A) measured at a spatial location corresponding to CRRES trajectory (Li et al., 1993).

quasi-periodic time variation of the inner belt electrons and protons (Blake et al., 1992; Li et al., 1993). Recently, the Van Allen Probes found a third belt in-between the outer belt and inner belt during a magnetic storm (Baker et al., 2013b; Mann et al., 2016).

Interestingly, similar radiation belt activities occur in other magnetized giant planets (Mauk and Fox, 2010). The understanding of terrestrial radiation belts therefore provides the basic principle for future human activities beyond the Moon and Mars toward more extreme radiation belts of giant planets (Table 1.1) which contain a variety of large moons (Table 5.1).

CHAPTER 3

Technological vulnerability and statistics

3.1 Effects of solar flares

Solar flares are electromagnetic wave bursts, which can disrupt radio communications on dayside Earth (Fig. 3.1). The flare impact starts instantaneously as it is electromagnetic radiation. Note also that it takes ~8 minutes for traveling from the Sun to the Earth by the speed of light (Fig. 2.1).

On December 12, 1901, Guglielmo Marconi successfully made the first transatlantic communication between Newfoundland and England using radio wave (Lanzerotti, 2017). Since Marconi's experiment, the ionospheric reflection of radio waves has been widely used for global communications. The radio waves below the *plasma frequency* f_p cannot propagate the plasma and are reflected:

$$f_p = \frac{1}{2\pi}\sqrt{\frac{N_e e^2}{m_e \varepsilon_0}}, \tag{3.1}$$

where ε_0 is the permittivity of free space, e is elementary charge, and m_e is electron mass. The easy-to-remember approximation is $f_p \sim 9.0\sqrt{N_e}$, where the electron density N_e is in a unit of m^{-3}. In the ionosphere, the plasma frequency is 2 to 20 MHz. The variable condition of the ionosphere therefore affects wide range of radio wave users.

If the electron–neutron collision frequency is higher than the radio wave frequency, the radio waves are absorbed. When the ionospheric D region appears in the dayside as solar flare X-rays suddenly increase, the HF (3–30 MHz) radio waves are therefore absorbed and the HF communication is disrupted, which is called the *Dellinger effect*.

In an extreme example, Knipp et al. (2016) summarized the first significant operational impact arising from major solar flares in May, 1967. The solar radio burst caused a direct radio interference at frequencies between 0.01 and 9.0 GHz, and also near-simultaneous disruptions of dayside radio communication, which might have had a critical societal impact under the situation of the Cold War.

Extreme Space Weather
DOI: https://doi.org/10.1016/B978-0-12-822537-0.00002-8

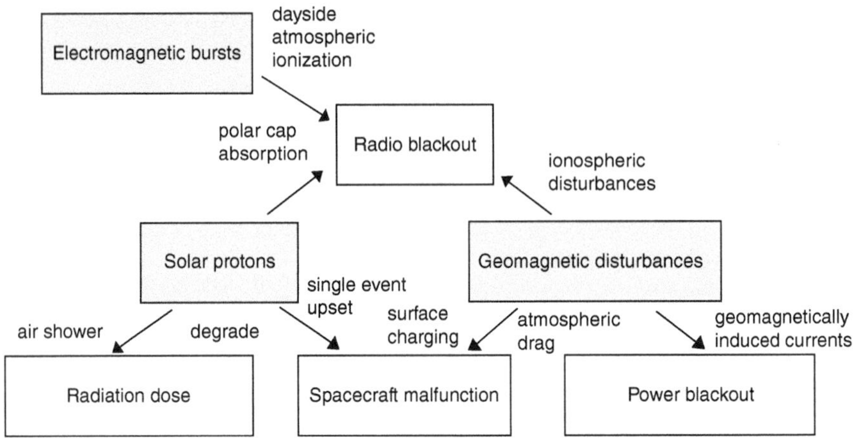

Fig. 3.1 ***Flow diagram from the disturbed space weather to possible space hazards.***

The modern society relies on the Global Navigation Satellite System (GNSS), which measures the position of a receiver using 1.2 or 1.5 GHz signals from multiple satellites. The errors of the measured position depend on the *total electron content* (TEC) because the group speed of the radio waves with higher frequency than f_p decreases in the plasma, as briefly shown below.

Refractive index of radio waves, $n_{ph} \equiv c/v_{ph}$, propagating in ionized medium by the phase speed v_{ph} is described in the well-known Appleton–Hartlee equation. By ignoring the collision and magnetic field, the Appleton–Hartlee equation can be simplified to:

$$n_{ph} \sim \sqrt{1 - \frac{f_p^2}{f^2}} \tag{3.2}$$

The group refractive index, $n_{gr} \equiv c/v_{gr}$, can be expressed via Eq. (3.1) as:

$$n_{gr} = n_{ph} + f\frac{dn_{ph}}{df} \sim 1 + \frac{40.3N_e}{f^2} > 1 \tag{3.3}$$

Thus, the group speed v_{gr} becomes smaller than the light speed, as the electron density increase.

Moreover, the GNSS satellites are sometimes unlocked by the irregularity of electron density pattern which interferes with the radio waves to cause a scintillation. Extreme effects from solar flares are summarized by NOAA/SWPC in Table 3.1. Some other ionospheric impacts are graphically summarized at the website of National Institute of Information and Communications Technology (Fig. 3.2).

Table 3.1 Possible extreme effects from solar flares (NOAA/SWPC).

Infrastructure	Effects from solar flares
HF Radio	Complete HF (high frequency) radio blackout on the entire sunlit side of the Earth lasting for a number of hours. This results in no HF radio contact with mariners and en route aviators in this sector.
Navigation	Low-frequency navigation signals used by maritime and general aviation systems experience outages on the sunlit side of the Earth for many hours, causing loss in positioning. Increased satellite navigation errors in positioning for several hours on the sunlit side of Earth, which may spread into the night side.

3.2 Effects of geomagnetic disturbances (GMD)

Among many engineered systems and infrastructures sensitive to extreme space weather (Cannon et al., 2013), one of the most vulnerable infrastructures is power grids (Fig. 3.3). The Hydro-Quebec blackout in March, 1989 is the most famous example (Allen et al., 1989). Six million Canadians lost electrical service for nearly 9 hours, with the estimated damage cost being

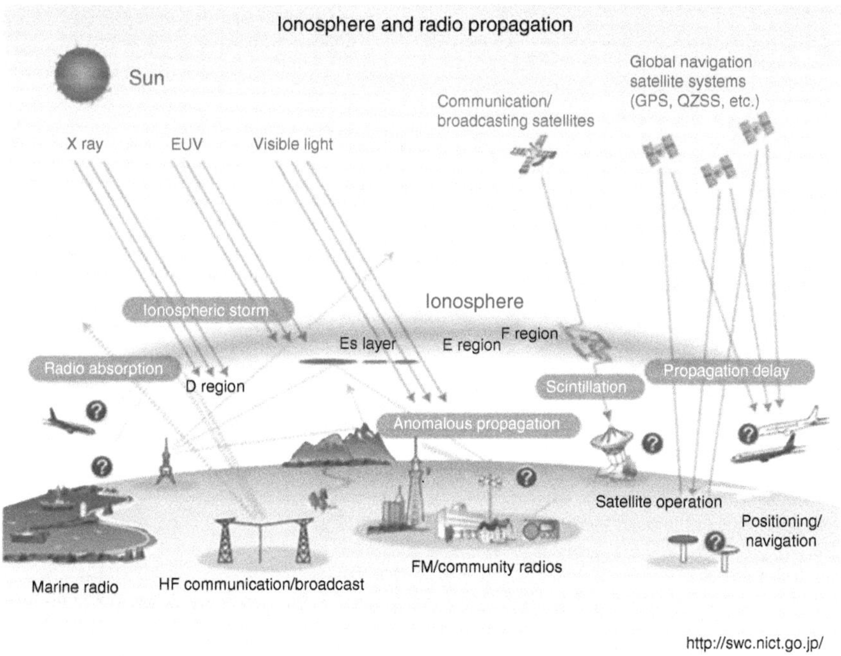

Fig. 3.2 ***Ionosphere and radio propagation (https://swc.nict.go.jp/en/knowledge/guide.html).*** *Copyright: National Institute of Information and Communications Technology (NICT), Japan.*

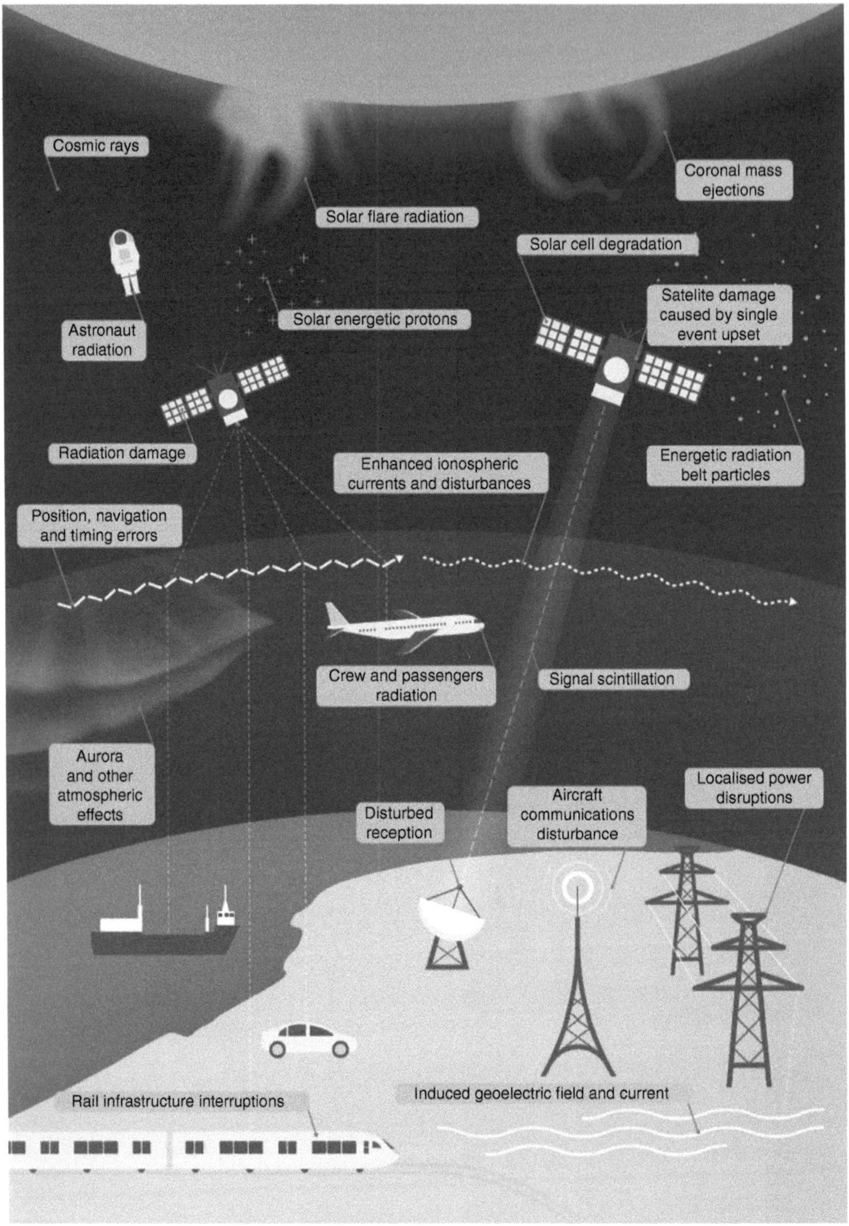

Fig. 3.3 ***Space weather phenomena and vulnerable infrastructures in modern society (UK Severe Space Weather Preparedness Strategy, 2021).***

$ 2 billion. It also permanently wiped out a $ 12-million transformer in the USA and two large UK transformers had to be sent back to the factory for repairs (Hapgood, 2012). The magnetic storm caused other distribution and equipment failures as well, including transformer failure at one nuclear generation plant. The time profile of the horizontal north–south component of the geomagnetic field at Kakioka, Japan is shown in Fig. 3.4.

Before the beginning of the storm main phase at 1100 UT, the second SSC occurred at 0743 UT on March 13, 1989. The start of trouble in Hydro-Quebec blackout was indicated by the first tripping of equipment at 0744:17 UT, and the final system collapse occurs at 0745:49 UT (Guillon et al., 2016; Boteler, 2019). In New Jersey, a power transformer was burned out due to the induced currents caused by the magnetic storm in March, 1989. Lakhina and Tsurutani (2016) noted that the March 1989 event is the largest magnetic storm occurred in the space age (minimum Dst index of −589 nT and minimum SYM-H index of −710 nT), and the main phase was complex and longer than that of usual storms. Other outstanding examples include the blackout event in Sweden during the Halloween storm in the end of October, 2003 (Pulkkinen et al., 2005).

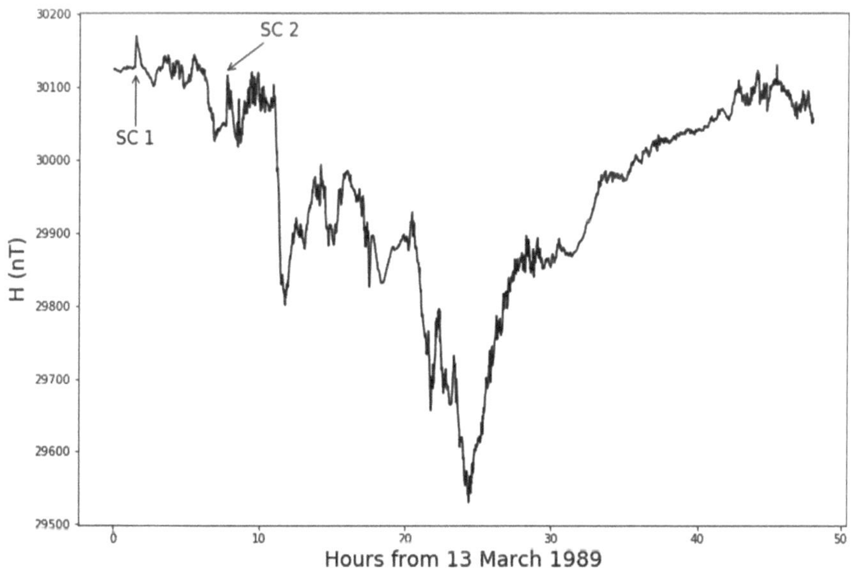

Fig. 3.4 ***Time profile of the north–south component of magnetic field at Kakioka, Japan during the magnetic storm in March, 1989.*** The start of trouble in Hydro-Quebec blackout was indicated by the first tripping of equipment at 0744:17 UT, ~1 minute after the second SC event.

GMD can therefore cause power outages at major cities. If another event like Carrington event in 1859 were to occur today, it could lead to an economic loss of US$ 1-2 trillion for USA (Baker et al., 2008). The mechanism behind is that GMD can induce abnormal electric current flow in an artificial conductor, which is called *geomagnetic-induced currents* (GIC).

GIC causes unwanted imbalances in system transformers that lead to long-term damage, bothersome audio noise, tripping of protective relays, and even catastrophic failure. In addition to the March 1989 accident, Neil Davis said in his book *The Aurora Watcher's Handbook* (1992, University of Alaska Press, Fairbanks, p157) that "An expensive 735 kV transformer at Hydro-Quebec's facility on James Bay, Canada, failed during a great red auroral display in 1980. Four months later, another major auroral display occurred and destroyed the transformer's replacement." It is notable that extensive auroral displays simultaneously occurred during the great magnetic storms. In Alaska, the oil pipeline also suffered from GIC (Campbell, 1980). The GIC risk is, however, not limited in polar regions. Ngwira et al. (2013b) noted that GIC around the *equatorial electrojet* can be an order of magnitude larger than the locations outside the equatorial electrojet. Therefore, power systems located around the equatorial electrojet can be exposed to increased GIC levels than those in the midlatitudes, which increases their susceptibility to GIC risk.

The basic physics to connect GMD and GIC is explained by Maxwell's law, that is, Eqs. (1.3) to (1.6). The Earth surface, called the crust, is a conductor. In case of GMD, Maxwell's law consistently completes the electric field distribution at the surface, which is called geoelectric field. If a transmission line connects the different points of the surface as shown in Fig. 3.5, the induced electromotive force drives the GIC in the transmission line. Assuming a homogeneous Earth, the skin depth is defined as the penetration depth of the electric field as:

$$d = \sqrt{\frac{1}{\mu \sigma \pi f}} \tag{3.4}$$

where σ is the conductivity in S/m, μ is the permeability of Earth in H/m and f the frequency in Hz. Assuming the permeability of the Earth is close to that of free space, that is, $\mu_0 = 4\pi \times 10^{-7}$, the periodicity of 100 seconds and the conductivity of 0.01 S/m give the skin depth of ~50 km.

GIC response to GMD therefore depends on the underground conductivity structure. For example, Pulkkinen et al. (2012) reported the one-in-100-year geoelectric field amplitudes are 20 and 5 V/km at high-latitude

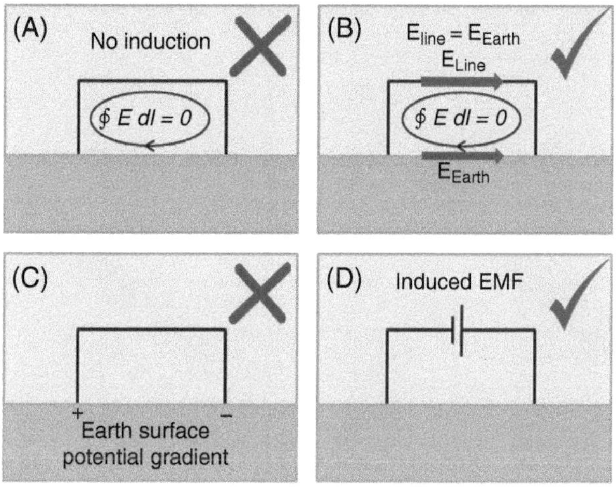

Fig. 3.5 ***Incorrect and correct views of geomagnetically induced currents (GIC).*** (A and B) The integral of electric field around loop formed by a power line and the Earth surface equals to 0 because the E_{line} equals E_{Earth}. (C and D) GICs are driven by the induced electromotive force in the transmission line, which acts like an extra voltage source (Boteler and Piryola, 2017).

locations with poorly conducting and well-conducting ground structures, respectively. As another example, the depth profile of the conductivity model in Japan is different from that in Finland, due to the subduction zone dynamics underneath Japan (Pulkkinen et al., 2010). GICs in regions with heterogeneous ground resistivity were discussed in a special issue of *Earth, Planets and Space* (Fujita et al., 2016).

From Faraday's law, Eq. (1.4), *magnetotelluric* (MT) plane-wave approach (Simpson and Bahr, 2005; Bonner and Schultz, 2017) is often used to estimate the underground structure. Using empirically derived surface impedance tensors Z, the horizontal geoelectric field $\mathbf{E}$ can be approximately computed from horizontal geomagnetic field $\mathbf{B}$ as follows.

$$\mathbf{E}(\omega) = \frac{1}{\mu}\begin{bmatrix} Z_{xx} & Z_{xy} \\ Z_{yx} & Z_{yy} \end{bmatrix}\mathbf{B}(\omega) \tag{3.5}$$

where μ is magnetic permeability, and all quantities are expressed in the frequency domain. Similar methods are used to obtain the possible values of geoelectric field and GIC amplitude in Japan during extreme magnetic storms, including the Carrington storm (Ebihara et al., 2021).

Kataoka and Pulkkinen (2008) showed that both sheath region and the CME body are important solar wind structures to drive large GIC events. The

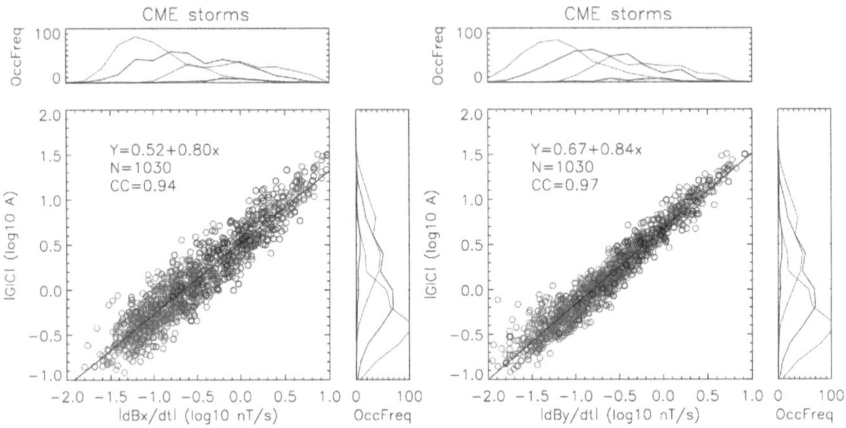

Fig. 3.6 ***Scatterplot and histograms of dB/dt against GIC during the different phase of magnetic storms (Kataoka and Pulkkinen, 2008).***

importance of CME sheath is also highlighted by Huttunen et al. (2008). The direct causes of the largest GIC events are associated with either SCs, substorms, or magnetic storms, although the relationship between dB/dt and GIC is always the same (Fig. 3.6). Note that SC, substorm-related auroral electrojet (AE index), and magnetic storm-related ring current (Dst index) have different time scales, ranging from seconds to hours (Kataoka and Ngwira, 2016). One of the largest SSC on March 24, 1991 caused the largest GICs ever measured in the USA (Kappenman, 2003).

Even during moderate Dst storms, super substorms (Tsurutani et al., 2015) can lead to severe GICs especially in mid to high latitudes. During the August 1972 event, telecommunication cable systems at Plano, Illinois in the USA was disrupted, with dB/dt amplitude of > 2000 nT/min (Anderson et al., 1974). It is also notable that the August 1972 event was the shortest transit event and was caused by a faster CME than that caused Carrington event. The Dst storm amplitude of this event was, however, moderate as shown in Table 2.7.

The earliest space-weather accident in Japan was observed on September 25, 1909, in which GIC affected the submarine cable connecting Tokyo and Guam (Uchida, 1909). The estimated −Dst amplitude is ~600 nT (Love et al., 2019a), which is comparable to that of March 1989 event. As shown in Fig. 3.7, the GIC in May 1921 affected the whole world, as summarized in Hapgood (2019). It is notable that the estimated −Dst amplitude is ~900 nT, which is comparable to that of the Carrington event. Love et al. (2019b) have summarized the problems on the New York

Railroad systems such as telephone lines, telegraph operations, and signal switching by the GIC during this super storm (Fig. 3.7).

Royal Academy of Engineering, UK (2013) recommended their stakeholders as follows: A) Government and industry should consider the potential for space weather damage on the optical fiber network through overvoltage on the repeaters and should consider whether appropriate assessment studies are necessary. B) UK railway operators and pipeline

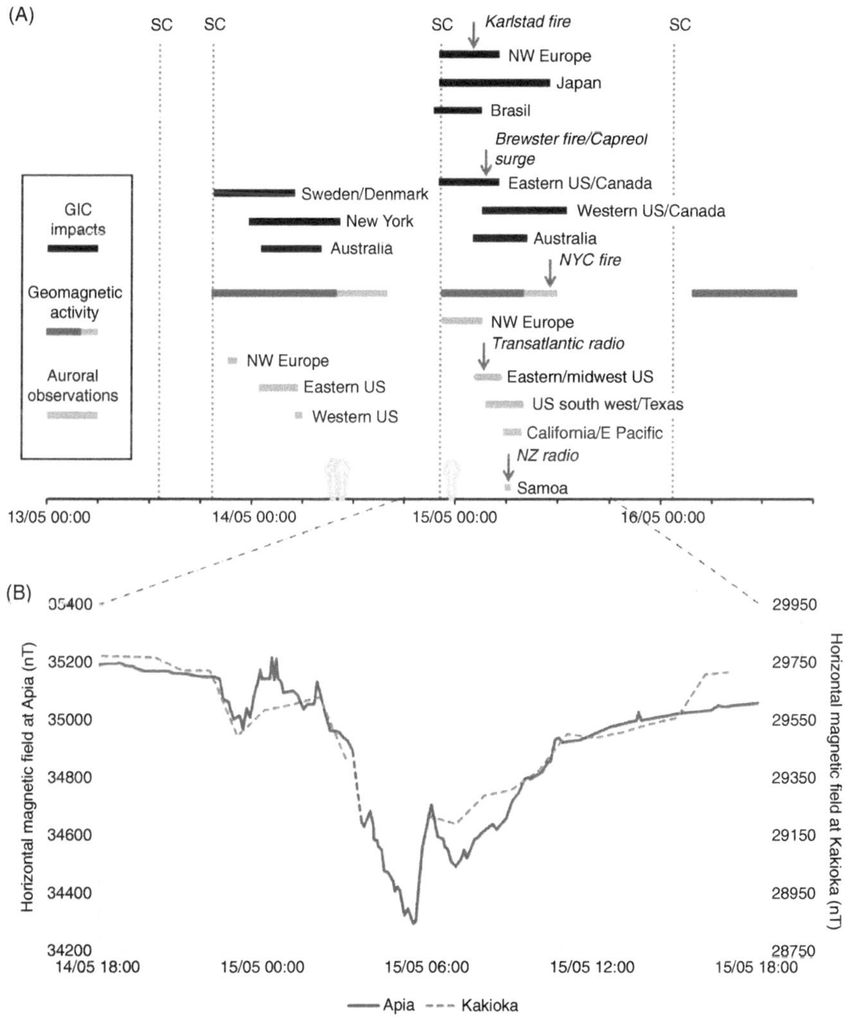

Fig. 3.7 ***Worldwide impacts of geomagnetically induced currents (GIC) during an extreme magnetic storm in May 1921 (Hapgood, 2019).***

operators should be briefed on the space weather and GIC risk and should consider whether appropriate assessment studies are necessary. Pulkkinen et al. (2017) have summarized the application readiness level as identified from the GIC Working Group of NASA Living with a Star Institute. The scientific outputs have also been introduced by Pulkkinen (2017).

GMD have potentially severe impacts other than those from GICs, as shown below. They are indirect effects, but the first example is due to the atmospheric density enhancement in the thermosphere (atmospheric drag during magnetic storms), the second example is due to the energetic electron enhancement in the magnetosphere (surface charging of satellites during substorms), and the last example is due to the electron density enhancement in the ionosphere (communication error during magnetic storms).

The large substorm-related electric currents flowing in the polar atmosphere cause the Joule heating of the upper atmosphere in the polar region, which then expands the atmosphere further upward to touch the LEO. The enhanced atmospheric drag affects the attitude control of the LEO satellites.

During the Bastille Day magnetic storm on July 15, 2000, the atmosphere sporadically expanded and the atmospheric density at the 440 km altitude of Japanese ASCA satellite suddenly increased to several times larger than the normal value. This caused an external torque on the satellite due to the increased air drag, which perturbed the satellite attitude. Accordingly, the on-board attitude control system made ASCA to the safe-hold mode. The solar paddles were then moved away from the nominal direction and the battery was exhausted, leading to the death of the satellite earlier than the expected reentry. As the latest example, during moderate magnetic storms on February 3-4, 2022, SpaceX Falcon 9 launched 49 Starlink satellites to low Earth orbit. Falcon 9's second stage deployed the satellites into their intended orbit, with a perigee of ~210 km. Notifying that the atmospheric drag was 50 percent higher than during previous launches due to the magnetic storm, the Starlink team commanded the satellites into a safe-mode where they would fly edge-on to minimize drag to effectively take cover from the storms. However, 38 Starlink satellites reentered the atmosphere, due to the unexpected space weather (Kataoka et al., 2022).

The large electric currents also flow in the magnetosphere during substorms. The current-carrying electrons and resultant radiation belt electrons cause the surface charging of the GEO satellites and this charge–discharge process causes satellite malfunctions. The NESC 2016 Technical Update (NP-2016-11-858-LaRC) reported an example of the Galaxy 15 satellite as follows: "In April 2010, the Galaxy 15 telecommunications satellite was set adrift, wandering away from its assigned place in geosynchronous orbit. Ground controllers had lost contact with the spacecraft and were powerless to upload the commands it needed for executing station keeping maneuvers. As a result, the satellite began an uncontrolled drift in longitude, with the threat of getting in the way of other satellites, and potentially interfering with their transmissions. This uncontrolled drifting went on for months. Reports in the scientific literature and space technology trade journals suggested Galaxy 15 was a victim of spacecraft charging. Hot electrons roaming the outer radiation belt had pelted the satellite, causing a negative charge to build on its surface—a charging event. And much like walking across a carpet and then touching a door knob, an electrostatic discharge ensued that knocked out its communications systems. It could no longer receive radio contact from its owner, Intelsat. It took 8 months to reestablish contact and successfully reposition the satellite in its desired orbit."

The modern high-tech society depends on GNSS systems. Ionospheric irregularities, such as those associated with *plasma bubbles*, directly cause radio scintillation, which adds noise on transmitted satellite images. The plasma bubble at equatorial region is basically caused by Rayleigh–Taylor instability and forms small-scale density irregularities. During magnetic storms, the bulk of equatorial plasma is lifted up by the enhanced electric field to a few thousand km and falls along the magnetic field lines to the middle latitudes, which is called *super-fountain effect* (Fig. 3.8). An extreme

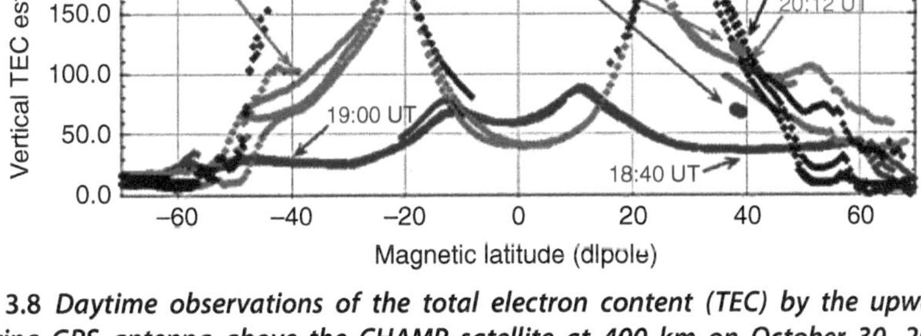

Fig. 3.8 ***Daytime observations of the total electron content (TEC) by the upward-viewing GPS antenna above the CHAMP satellite at 400 km on October 30, 2003.*** Drastic TEC enhancement was observed after the beginning of the magnetic storm at ~2000 UT (Mannucci et al., 2005).

Table 3.2 Possible extreme effects of geomagnetic disturbances (GMD) (NOAA/SWPC).

Infrastructures	Effects from geomagnetic disturbances
Power systems	Widespread voltage control problems and protective system problems can occur; some grid systems may experience complete collapse or blackouts. Transformers may experience damage.
Spacecraft operations	May experience extensive surface charging, problems with orientation, uplink/downlink, and tracking satellites.
Other systems	Pipeline currents can reach hundreds of amps, HF (high frequency) radio propagation may be impossible in many areas for one to two days, satellite navigation may be degraded for days, low-frequency radio navigation can be out for hours, and aurora has been seen as low as Florida and southern Texas (typically 40° geomagnetic latitude).

consequence of the equatorial ionization anomaly during Carrington-class magnetic storm has been discussed by Tsurutani et al. (2012). The enhanced density increases navigation error, while the irregularities unlock GNSS satellites, both of which make the GNSS system unavailable.

During magnetic storms, associated with the enhanced ionospheric convection, storm-enhanced density (SED) events also occur in middle latitude at afternoon sector. Further, there is an interesting example that electron and ion temperatures of the topside ionosphere increased several hundred K, but NmF2 significantly decreased from those in quiet time (Liu et al., 2016). Note also that super-fountain-like effects do not necessarily occur in the presence of strong dawn-to-dusk electric field (Rajesh et al., 2021). It is likely that similar effects are more frequent than previously expected, although a knowledge gap still remains.

The possible effects due to large-amplitude GMD are briefly summarized in Table 3.2.

3.3 Effects of energetic particles

We are flying high beyond the radiation shelter of the atmosphere to be susceptible to energetic particles. GCR and SEP form the air shower in the stratosphere, which include highly penetrating particles such as neutrons and muons. At a flight altitude, typically at 12 km altitude, the energy deposition of these particles causes the radiation dose in the body of air crews. Protons of >400 MeV can contribute to such a radiation dose at flight altitude. The SI unit of absorbed dose is Gy = J/kg, while the dose equivalent indicates the biological risk, and the unit is differentiated as Sv.

As the worst case, a total dose at typical flight altitude in the polar region for the largest single GLE event is estimated to be 0.33 mSv (Facius and Reitz, 2007). This happened on February 23, 1956 (GLE 5), which is assumed to be 10 times the dose from the GLE event on 29, September 1989 (GLE 42).

During *extravehicular activities* (EVA) at ISS, the amount of radiation dose is severe, because a large portion of the primary protons of SEP penetrates the spacesuit. Even inside the spacecraft, the radiation dose can be severe, because a part of primary protons penetrates the spacecraft wall. On October 19, 1989 (GLE 43), there was an action to avoid radiation dose from SEP at the most shielded space inside ISS. The measured dose during GLE 43 at MIR was 27.0 mGy, which is more than several times larger than that of the other GLE events in 1989: GLE 42 (4.9 mGy), GLE 44 (2.9 mGy), and GLE 45 (1.4) (Shurshakov et al., 1999). It is noteworthy here that it was only 0.1 mGy at ISS and 0.4 mGy at Mars surface during the 10 September 2017 event (Berger et al., 2018), probably due to the grasping CME impact (Kataoka et al., 2018). An example of the radiation dose rates at ISS due to SEP, GCR, outer belt, and SAA during disturbed space weather periods in June, 2015 were introduced by Dachev et al. (2017).

The famous "light flash" phenomenon, first reported by Buzz Aldrin during Apollo 11 mission in 1969, was caused by the energetic ions of GCR (Fazio et al., 1970). Similar light flashes have been continuously observed by many astronauts, even at ISS. Such radiation damages are not limited to humans as shown below.

The CCD, solar panels, and many other electronic devices equipped onboard spacecraft also suffer from SEP radiation. Fig. 3.9 shows the example images of coronagraph during an SEP event. The scratched pattern or "snow storm" of the images is caused by the energetic ions. Sensitive instruments are often contaminated by SEP, which causes data gap. Permanent solar panel degradation due to SEP radiation is critical for spacecraft because it directly shortens their life. When a single energetic particle causes malfunction in common on-board electronics, the response produced is called *single event effect* (SEE). SEE can take on many forms. Among them, bit inversion due to the passage of an energetic particle, that is, nondestructive soft errors, can cause serious malfunctioning of computers onboard the spacecraft, which is called *single event upset* (SEU).

As an example, the SEP events on September 29 and October 19, 1989 degraded the solar panel onboard Japanese meteorological satellite

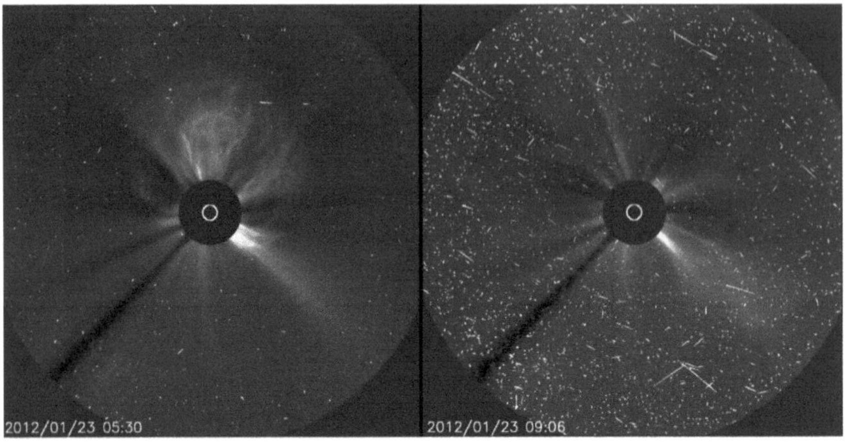

Fig 3.9 ***So-called "snowstorm" effect on CCD camera as observed by SOHO/LASCO coronagraph.*** A major M8.7 solar flare and a fast halo CME began early on January 23, 2012, igniting the most intense SEP event since 2005. Interestingly, only a minor magnetic storm (peak Dst index = −71 nT) occurred afterward, mainly because of the lack of CME-associated SBZ during the passage of the Earth.

Himawari-3 at GEO, which was equivalent for 3-year degradation at once. The other satellites simultaneously reported the increase in the number of SEU events. As another example, solar proton events in March 1991 degraded the solar panel onboard Japanese broadcasting satellite Yuri-3a and the three-channel broadcast was disabled. At the same time, SEU events are also frequent at GEO (Shea et al., 1992), as well as in the slot region by the unexpected intrusion of SEP particles (Mullen et al., 1991).

The atmospheric ionization causes some more effects in polar regions. The large portion of the solar protons enters the open magnetic field region called polar cap to form the D region. The D region formation has an effect similar to Delinger effect, called *polar cap absorption* (PCA), to cause a radio wave communication failure. The solar protons in the polar cap also destroy the stratospheric ozone, as discussed in Chapter 2. Interestingly, the mesospheric ionization during the PCA events can cause molecular nitrogen emission at 391.4 nm, known as polar glow aurora (Sandford, 1961, 1963).

The possible effects of large-amplitude SEP events are summarized by NOAA/SWPC in Table 3.3.

Solar protons have been mainly discussed above, but it is also meaningful to discuss another radiation effect due to electrons in the remaining of this Section. It is hard to stop electrons than ions. Hence, radiation belt electrons are called *killer electrons* and cause the problems to safely operate the

Table 3.3 Possible extreme effects from solar energetic particles (SEP) (NOAA/SWPC).

Infrastructure	Effects from solar energetic particles
Biological	Unavoidable high radiation hazard to astronauts on EVA (extra-vehicular activity); passengers and crew in high-flying aircraft at high latitudes may be exposed to radiation risk.
Satellite operations	Satellites may be rendered useless; memory impacts can cause loss of control; may cause serious noise in image data; star-trackers may be unable to locate sources; permanent damage to solar panels possible.
Other systems	Complete blackout of HF (high frequency) communications possible through the polar regions, and position errors make navigation operations extremely difficult.

satellites (Baker and Lanzerotti, 2016). *Deep dielectric charging* occurs when high-energy electrons penetrate satellite shielding materials. The physics of the internal charging (spatial and time variation of charge density ρ) can be described by Poisson's equation as follows:

$$\nabla \cdot E = -\nabla^2 \Phi = \frac{\rho}{\kappa \varepsilon_0} \tag{3.6}$$

$$\frac{\partial \rho}{\partial t} = -\nabla \cdot (J_R + J_C), \tag{3.7}$$

where κ is the dielectric constant of the material, J_R is radiation current, and J_C is conduction current. Evaluation of internal charging therefore requires solving both radiation transport to compute the charge deposition within the material and electrostatic potentials due to the accumulating charge density (Lai et al., 2018).

As an extreme example, three communication satellites ANIK-E1, E2, as well as Intelsat-k at geosynchronous orbit stopped their operation simultaneously on January 20, 1994 due to the momentum wheel control malfunction. This is so-called "phantom command" which was not manually commanded from the ground control station. Although E1 recovered in a few hours, E2 recovery took 6 months and cost \$ 50 to 70 million (National Research Council, 2008). The deep dielectric charging and discharging likely caused this event because there were no solar flare activities. Moreover, the high radiation belt electron flux continued for a week (Wrenn, 1995). Some other possible impacts in the radiation belts are summarized in Fig. 3.10. Many examples of space radiation effects on the space-based systems are summarized by Baker and Bodeau (2021) and Bodeau and Baker (2021).

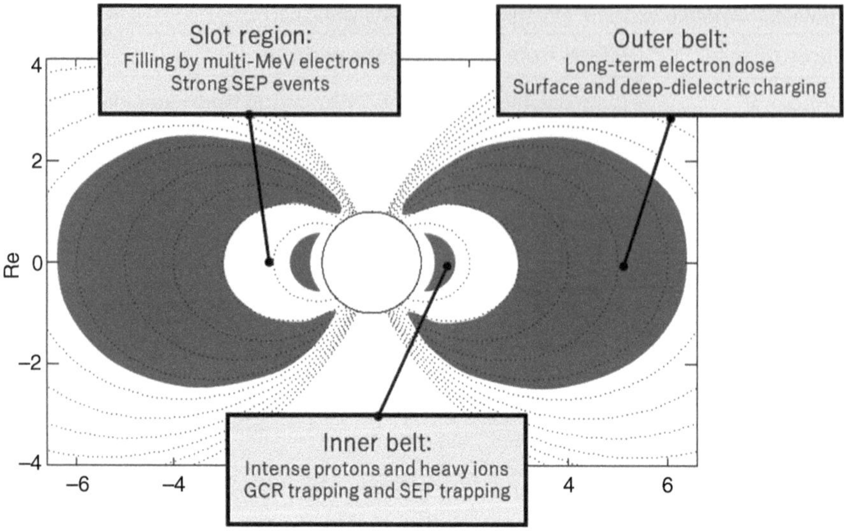

Fig. 3.10 ***Schematic illustration of the radiation belts and possible impacts.*** See also Baker et al. (2018).

The most extreme conditions of radiation belts were artificially induced by nuclear tests, starting from Starfish Prime exploded at 400 km altitude in July 1964. The inner belt was filled by the relativistic electrons of at least million times larger flux than usual. Several satellites were in malfunctional for a few years after the nuclear tests (Baker and Panasyuk, 2017).

3.4 Statistics of extreme events

The occurrence of magnetic storms is not a random process. As naturally expected, the peak sunspot number of a solar cycle influences the number of magnetic storms in the solar cycle (Kataoka, 2013). They are different in the odd and even cycles and the rising and declining phases of the solar cycle (Owens et al., 2021). In a long term, for example, low-latitude aurora sightings are rarely found in the grand minimum because they are related to large magnetic storms (Kataoka et al., 2017). However, note also that moderate solar cycles can also have extreme magnetic storms (Kilpua et al., 2015a).

The NOAA scales (Tables 2.1, 2.2, and 2.5) have the average probability per cycle as obtained in the recent space age. From the statistical comparison among several space weather phenomena of magnetic storms, solar flares, and CMEs, Riley (2012) estimated the probability of another

Carrington event in the next 10 years to be 12%. Kataoka (2013) applied the same analysis to the 90-year list of magnetic storms and estimated the probability of another Carrington storm in next 10 years to be 4 to 6%.

A useful concept to define the extreme events is so-called one in 100-year event. Tsubouchi and Omura (2007) applied an extreme value theory to estimate the amplitude of the 100-year event as −Dst = 550–750 nT. Love et al. (2015) also estimated the amplitude of 100-year event as −Dst = 880 nT, so that the Carrington event roughly corresponds to the 100-year event. The best efforts have been repeatedly conducted to estimate the amplitude of 100-year and even 1000-year event by carefully extrapolating the tail distributions, although such results are highly uncertain because of the limited data set of the Dst index available for only ~60 years (Love, 2020).

The similar statistical analysis can be applied for substorms as well. It has been discussed for a long time that the amplitude of substorms basically follows log-normal distributions (e.g., Liou et al., 2018). From the statistical analysis, Nakamura et al. (2015) estimated the possible maximum amplitude of substorms as AE ~ 4100 nT. In fact, super substorms close to the 4×10^3 nT were observed (Tsurutani et al., 2015). It is also impressive that a negative spike of 3700 nT in the north–south component was reported on 13 July, 1982 at high latitude (Viljanen et al., 2014).

In general, interactions among many elements or their nonlinearity produce a new characteristic from the complex network. In the complex system, large-amplitude events tend to follow a power-law distribution (Riley, 2012). The power-law distribution of the event amplitude x is defined as

$$f(x) = Ax^{-\alpha}, \tag{3.8}$$

where α denotes the spectral index and A is a constant. A useful way to investigate the large-amplitude and/or rare events is a cumulative distribution function which is defined as

$$N(x) = N_T \int_x^{\infty} f(x')dx' = \frac{N_T A}{\alpha - 1} x^{-\alpha+1} \tag{3.9}$$

The spectral index can be obtained by the maximum likelihood estimation (Newman, 2005):

$$\alpha - 1 = N_T \left\{ \sum_{i=1}^{N_T} \ln\left(\frac{x_i}{x_{\min}}\right) \right\}^{-1}, \tag{3.10}$$

where the $x_{\min}$ is the minimum value to be used for the fitting.

Rare or large-amplitude events arise from the majority, which usually follow a log-normal distribution in a complex system because it is characterized as a random walk of multiplications rather than that of additions.

The log-normal distribution is defined as

$$f(x) = \frac{1}{\sqrt{2\pi}\sigma x} \exp\left(\frac{(\ln x - \mu)^2}{2\sigma^2}\right), \tag{3.11}$$

where μ is the geometric mean value and σ is the standard deviation. The cumulative distribution function of the log-normal distribution can be written as

$$N(x) = N_T \int_x^\infty f(x')dx' = \frac{N_T}{2}\left\{1 - \operatorname{erf}\left(\frac{\ln x - \mu}{\sqrt{2}\sigma}\right)\right\}, \tag{3.12}$$

where N_T is the total number of events and the error function is

$$\operatorname{erf}(x) = \frac{2}{\sqrt{\pi}} \int_x^\infty \exp(-y^2)dy \tag{3.13}$$

To estimate the possible amplitudes of the 100- or 1000-year events from a limited data set of less than 100 years, both power-law distribution (Eq. 3.8) and log-normal distribution (Eq. 3.11) can be fitted to the data set. The time-stationarity of the statistical distributions is therefore assumed. The log-normal distribution gives relatively conservative estimates, while the power-law distribution generally gives upper-limit estimates (Riley and Love, 2017).

Due to the long-term monitoring of geomagnetic activities at Kakioka Magnetic Observatory, Japan, their complied lists of magnetic storms and SCs can be used for statistical estimation of the possible amplitudes of the 100- and 1000-year events. The possible power-law distributions of the amplitude of magnetic storms as well as SCs were discussed by Minamoto et al. (2015), using the event list. It should be noted that the amplitude of magnetic storms is ranked by the local magnetic field variation in the horizontal component dH, and not a global average as given from the Dst index.

Fig. 3.11 shows that both log-normal and power-law distributions can be fitted to the large-amplitude events of magnetic storms. Conservative estimates by the log-normal fitting suggest the 100-year and 1000-year amplitudes to be 750 and 1200 nT, respectively.

With regard to the possible errors for similar log-normal fitting methods applied to the Dst index, it was estimated that a Carrington-class magnetic storm with −Dst >850 nT occurs about 1.13 times per century and a wide 95% confidence interval of 0.42 to 2.41 times per century, while a 100-year magnetic storm has −Dst > 880 nT, but a wide 95% confidence interval of 490 to 1187 nT (Love et al., 2015).

Fig. 3.12 shows that log-normal distribution fits the small-amplitude SC events, while power-law fits the large-amplitude SC events. The amplitude

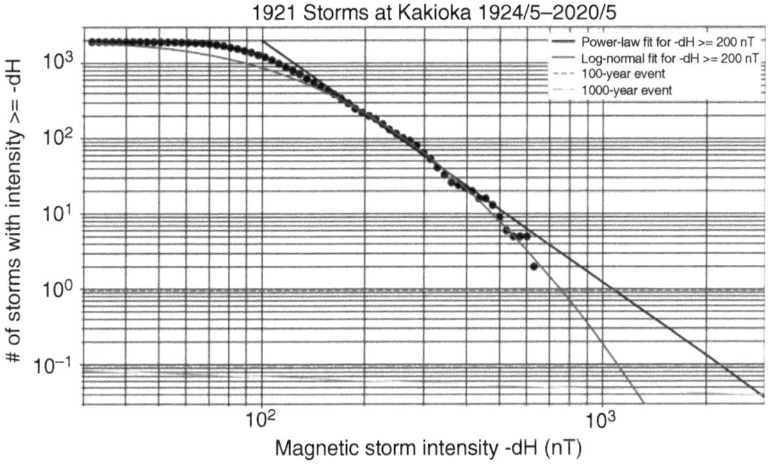

Fig. 3.11 ***Cumulative distribution function of magnetic storms in dH amplitude, as observed at Kakioka Magnetic Observatory, Japan.*** The fitted curves are log-normal distribution and power-law distribution (Kataoka, 2020).

of the 100- and 1000-year SC events are therefore estimated from the power-law fit to be 200 and 400 nT, respectively.

The substorm list and SME index, substitute of AE index, are available at SuperMAG website (https://supermag.jhuapl.edu/). The obtained cumulative distribution function of substorms follows log-normal distributions

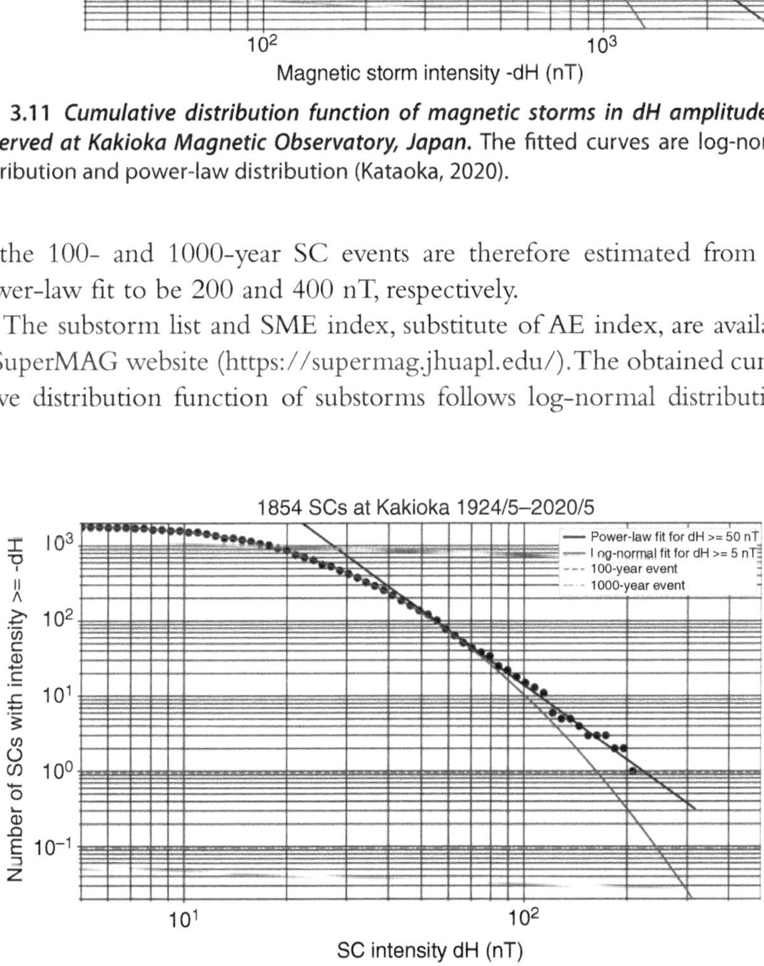

Fig. 3.12 ***Same as Fig. 3.11, except for the amplitude of sudden commencements (SCs) (Kataoka, 2020).***

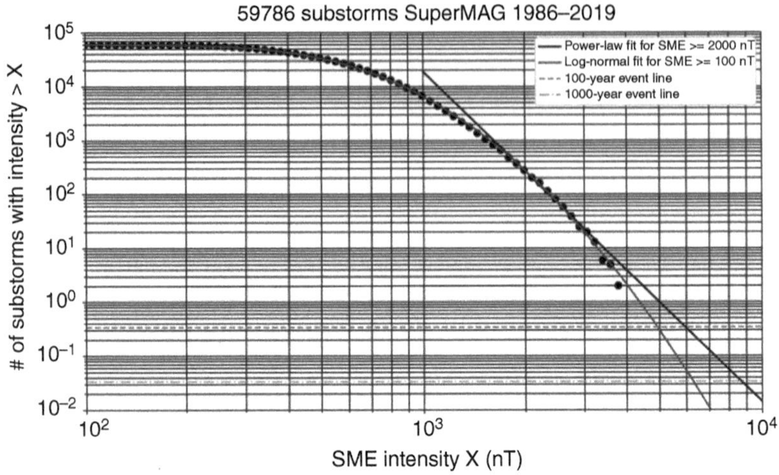

Fig. 3.13 ***Cumulative distribution function of substorms as identified from the SuperMAG project.*** It is shown that substorms basically follow a log-normal distribution (Kataoka, 2020).

(Fig. 3.13). The amplitudes of the 100- and 1000-year substorm events can be estimated as ~5000 and ~6000 nT, respectively. The estimated amplitudes of the different kinds of GMD (Kataoka, 2020) are summarized in Table 3.4.

The same statistical analysis can be applied to solar flares, CME speed, and SEP intensity. The solar flare event is available at NOAA website (https://www.ngdc.noaa.gov/stp/solar/solarflares.html). The CME event list is available at NASA website (https://cdaw.gsfc.nasa.gov/CME_list/). The SEP event list is available at NOAA and NASA (https://umbra.nascom.nasa.gov/SEP/).

Figs. 3.14, 3.15, and 3.16 show that the most solar flares follow power-law distribution instead of the log-normal distribution, while most SEP intensities and CME speeds follow log-normal distribution. The difference of cumulative distribution functions is probably because only specific solar flares produce SEP or CMEs (Kataoka, 2020).

Table 3.4 Possible amplitudes of 100- and 1000-year magnetic storm, sudden commencement (SC), and substorm, as estimated from log-normal or power-law fits.

	Magnetic storm (Dst)	SC (SYM-H)	Substorm (AE)
100-year event	750*~1100 nT	160~230* nT	5000*~6000 nT
1000-year event	1100*~2200 nT	240~450* nT	6200*~8500 nT

Asterisks denote the probable values from the better fitted distributions (Kataoka, 2020).

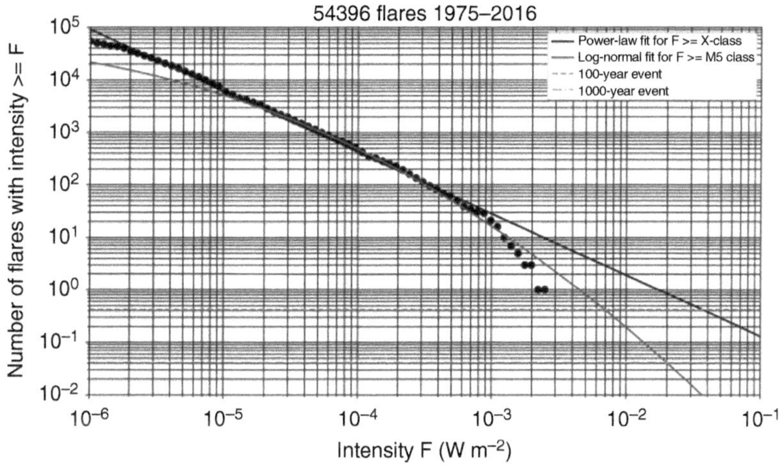

Fig. 3.14 ***Cumulative distribution function of solar flare intensity.*** Flare intensity follows the power-law distribution, except for the large-amplitude events (Kataoka, 2020). C, M, and X flares are 10^{-6}, 10^{-5}, and 10^{-4} W/m^2.

To summarize, the estimated amplitudes of 100- and 1000-year events, including those by Gopalswamy (2018), are summarized in Table 3.5. The difference is the types of distributions fitted to the cumulative distribution function. Goplaswamy (2018) used the Weibull distribution (exponential curve), while Kataoka (2020) uses the

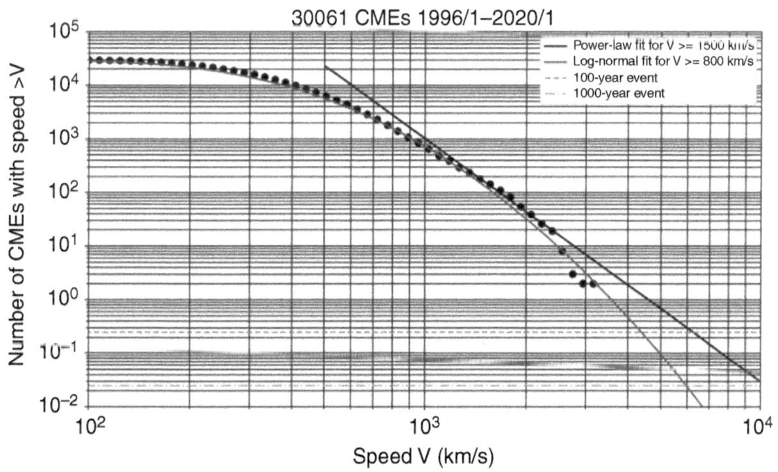

Fig. 3.15 ***Cumulative distribution function of CME speed (Kataoka, 2020).***

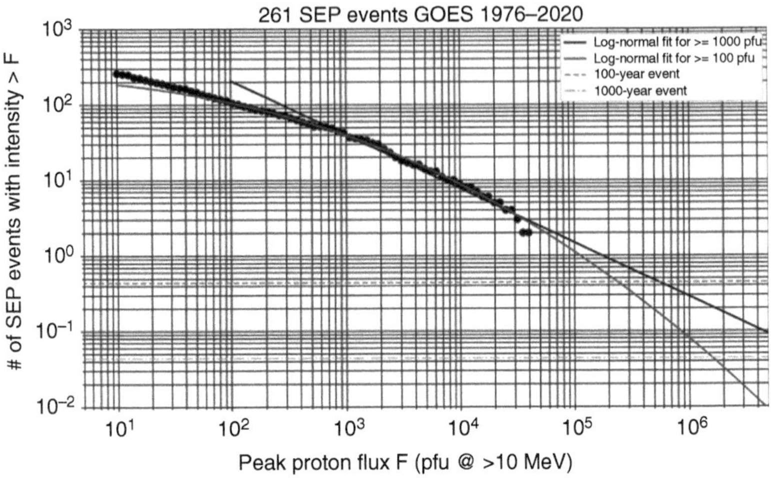

Fig. 3.16 ***Cumulative distribution function of solar energetic particle (SEP) intensities (Kataoka, 2020).***

log-normal and power-law distributions to give larger estimates than those of Gopalswamy (2018).

Similar 100-year event analyses are applied to the other phenomena. For example, the 100-year event of daily averaged >2 MeV electron fluxes at GEO is 3.2 to 7.7×10^5 cm^{-2} s^{-1} sr^{-1}, based on the estimation from 19.5-year GOES observation data since 1995 (Meredith et al., 2015). Kataoka and Nakano (2021) estimated that 100-year SBZ event has 137 nT. Jiggens et al. (2014) estimated the 100-year events of total ionizing dose, nonionizing energy loss, and SEU rate, as a function of the shielding thickness. Ishii et al. (2021) have reviewed the possible occurrence rates of extreme events for various types of space weather. Hapgood et al. (2016) have summarized other estimates of the worst space weather environments, which include the possible risks for power grids, satellite operations, electronics, wireless systems, GNSS system, railway system, and aviation. The possible economic impact of these possible effects still remains uncertain

Table 3.5 Estimated amplitudes of 100- and 1000-year flares, coronal mass ejection (CME) speeds, and solar energetic particle (SEP) intensities from Figs. 3.14–3.16 (Kataoka, 2020).

	Flares	CME speed	SEP
100-year event	X40–X70	3800–4500 km/s	1.5–2.5 10^5 PFU
1000-year event	X100–X200	4700–6000 km/s	1.0–1.5 10^6 PFU

The low estimates are from Gopalswamy (2018).

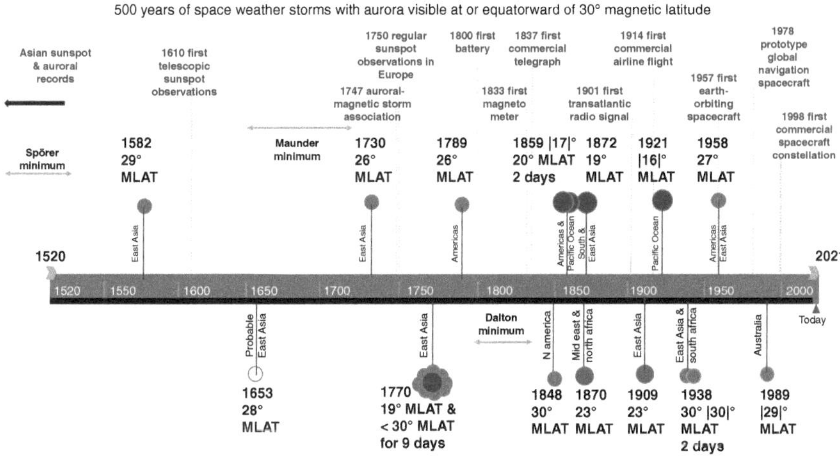

Fig. 3.17 ***Timeline of historical auroral events.*** Large circles correspond to equatorward auroral visibility (Knipp et al., 2021).

(Eastwood et al., 2017), but the knowledge obtained from these statistical distributions is useful to evaluate the technological vulnerability and to design a robust infrastructure and the safety criteria.

3.5 Outstanding space weather events

It is important for us to realize that "no two storms are alike" (Lanzerotti, 1992). Extreme space weather has many factors, as one can have extreme SEP event but not significant GMD, and vice versa. Outstanding space weather events in the last 500 years can be summarized by a timeline along the technology developments (Fig. 3.17). For example, it is evident from the timeline that September 1770 event (Kataoka and Iwahashi, 2017; Ebihara et al., 2017) was the special auroral event in 500 years. Modern outstanding space weather events after the Carrington event, as repeatedly introduced in literatures, can be briefly summarized below.

3.5.1 September 1–2, 1859: The Carrington event

The historically largest extreme magnetic storm occurred on the Earth (Tsurutani et al., 2003), 17.6 hours after the white light solar flare (~X45, Cliver and Dietrich, 2013). This Carrington event was the second fastest transit event (Cliver and Svalgaard, 2004). Discussions on the magnetometer record are countless (Akasofu and Kamide, 2005; Siscoe et al., 2006; Cliver and Dietrich, 2013; Cid et al., 2014, 2015). Many different models have reproduced the spiky variation with a rapid recovery (Li et al., 2006; Keika et al., 2015). The magnetic storm affected the worldwide telegraph system (Boteler, 2006) and auroras appeared worldwide (Kimball, 1960; Green and Boardsen, 2006; Hayakawa et al., 2016). Shea and Smart (2006) compiled the articles published in 1859–1861 on the Carrington event.

3.5.2 May 14, 1921: The New York Railroad event

Another Carrington-class magnetic storm (−Dst peak ~900 nT) causing the technological impacts on May 1921 (Kunitomi, 1921; Argenheister and Westland, 1921a, 1921b, Hapgood, 2019). Love et al. (2019b) have also summarized the confusions on the New York Railroad systems such as telephone lines, telegraph operations, and signal switching by the GIC during this super storm. According to the study by the Metatech Corporation, the occurrence of an event like the 1921 storm would result in large-scale blackouts affecting more than 130 million people and expose more than 350 transformers to the risk of permanent damage (National Research Council, 2008). As a spectacular example, telephone exchange in central Sweden was damaged by a fire started by the GIC (Karsberg et al., 1959). Silverman and Cliver (2001) summarized the worldwide auroral witness records, including the remarkable one from Apia, Samoa, within 13° of geomagnetic equator.

3.5.3 February 23, 1956: The greatest GLE event

The worst GLE event occurred in February 1956, just before the space age. The relative amplitude reached 9000% (Duggal, 1979), and the spectral hardness was distinguished. The GLE observation was made by ground-based instruments (Gold and Palmer, 1956; Marsden et al., 1956; Rishbeth et al., 2009). It can generate SEU rates greater than 1 per minute at 10 km altitude and 1 GeV rigidity cutoff (Dyer and Lei, 2001). Despite of its short duration, the fluence or total radiation dose was expected to be 0.33 mSv (Facious and Reitz, 2007).

3.5.4 February 11, 1958: The Tokyo aurora event

One of the greatest magnetic storms occurred on February 11, 1958. The Kp value reached the maximum value of 9, and the Dst peak was −426 nT (Table 2.7). The outstanding nature of this event was the large amount of red auroral witness reports worldwide (Winckler et al., 1959; Akasofu and Chapman, 1962). This was the latest event when we could see auroras from Tokyo, Japan near the north horizen. For a few hours from 1000 UT on 11 February 1958, northern lights appeared over northern Japan. The aurora was photographed for the first time in Japan, and several hand-made sketches were also recorded at the same time. This event therefore serves as a valuable point of contact between the modern recording of data and hand-written records (Kataoka et al., 2019; Kataoka and Kazama, 2019).

3.5.5 August 4, 1972: The high-speed and worst SEP event

It takes only 14.6 hours from the flare to SC on August 4, 1972 (Vaisverg and Zastenker, 1976). Cliver et al. (1990) discussed that the maximum solar wind speed at 1 AU would be ~2000 km/s. The record nearest magnetopause was observed at 5.2 Re (Hoffman et al., 1975). Notably, as shown in Table 2.6, the Dst index of this magnetic storm event reached only −118 nT, while the Kp index was record high. Cannon et al. (2013) discussed that this event was likely similar to the Carrington event, but with a northward IMF. Telecommunication was disrupted in the USA (Anderson, 1974). The August 4, 1972 event is also known as the worst SEP event, with an unacceptable risk if astronauts were on prolonged EVA (Jiggens et al., 2014) or walking on the Moon (Townsend, 2020).

3.5.6 March 13–14, 1989: The Hydro-Quebec blackout event

The largest Dst magnetic storm since 1957 occurred on March 13–14, 1989 (−Dst = 589 nT, −Sym-H = 710 nT, Lakhina and Tsurutani, 2016; Nagatsuma et al., 2015), and the equatorward auroral extent was ~40° (Allen et al., 1989; Boteler, 2019). The electric power industry has a good awareness of the geomagnetic threat after this power blackout event in Hydro-Quebec (Bolduc, 2002) as well as transformer damage in the UK (Erinmetz et al., 2002). Accordingly, the power grids have hardened their systems by installing transformers more resilient to power fluctuations and having enough reserve generators (Hapgood, 2012).

3.5.7 September–October, 1989: The MIR event

Solar proton events of September and October, 1989 are from the same active region with different solar rotation numbers. The latter had the largest flux SEP event, while the former had a hard spectrum (Jiggens et al., 2014). On October 19, 1989, during the GLE 32, special action was taken to avoid the significant SEP dose (~27 mGy) at MIR (Shurshakov et al., 1999). The SEP event degraded the solar panel of Himawari 3 at GEO, which is equivalent to 3 years degradation. Bright red aurora was observed from Hokkaido, Japan on October 21, 1989 (Miyaoka et al., 1990).

3.5.8 March 24, 1991: The super SC event

Approximately 3 hours after an impulsive solar flare of X9.4, a long-duration flare and large-amplitude SEP event occurred on March 23, 1991 (Shea et al., 1992). Anomalously large and short-duration SC was then observed on March 24, 1991 (Araki et al., 1997). The inner radiation belt was disturbed and "double-peaked" with the SC event (Mullen et al., 1991; Li et al., 1993). The largest-amplitude GIC was also observed over the USA associated with the SC event (Kappenman 2003). SEU events was frequent at GEO (Shea et al., 1992) and even in the slot region (Mullen et al., 1991). The solar panel of Yuri-3a broadcast satellite at GEO was degraded by the SEP event, and the full-channel broadcast was disabled.

3.5.9 July 15, 2000: The Bastille Day event

The Bastille Day event is characterized by a large SEP flux (24,000 pfu) and a fast average transit speed of ~1500 km/s (Watari et al., 2001). The solar flare of X5.7 was not extremely large, but the Kp index reached the maximum value of 9. Lanzerotti et al. (2001) reported the large trans-Atlantic voltage. During the magnetic storm, the Japanese ASCA satellite suffered from the suddenly increased air drag, and the solar paddles moved away from the nominal direction. The battery was exhausted since then, leading to the death of the ASCA satellite earlier than the expected reentry. A cutting-edge simulation study has reproduced this event (Torok et al., 2018).

3.5.10 October–November, 2003: The Halloween event

Successive CMEs of the Halloween event in 2003 (Gopalswamy et al., 2005; Shiota and Kataoka, 2016) caused the extreme effects including GIC (Pulkkinen et al., 2005) and super-fountain effect (Mannucci et al., 2005). The wide-ranging effects of the Halloween event are described as follows (NOAA, 2014): "The Sydkraft utility group in Sweden reported that strong GIC over Northern Europe caused transformer problems and even a system failure and subsequent blackouts. Radiation storm levels were high enough to prompt NASA officials to issue a flight directive to the ISS astronauts to take precautionary shelter. Airlines took unprecedented actions in their high latitude routes to avoid the high radiation levels and communication blackouts areas. Rerouted flights cost airlines $10,000 to $100,000 per flight. Numerous anomalies were reported by deep space missions and by satellites at all orbits. GSFC Space Science Mission Operations Team indicated that approximately 59% of the Earth and Space science missions were impacted. The storms are suspected to have caused the loss of

$640 million ADEOS-2 spacecraft. On board the ADEOS-2 was the $150 million NASA SeaWinds instrument. Due to the variety of this solar activity outbreak, most industries vulnerable to space weather experienced some degree of impact to their operations."

These events reminded both scientists and policy makers how significantly the space environment can affect human society and its various space- and ground-based technologies. National Research Council (2008) also added that GPS-based Wide Area Augmentation System (WAAS) implemented by the Federal Aviation Administrations was disabled for 30 hours. One special feature of the Halloween event was the occurrence of significant space weather effects on space missions away from the Earth, including the failure of the radiation monitor on NASA Mars Odyssey mission (Crosby, 2007) and severe interference in the star trackers on ESA Mars Express mission (Hapgood, 2011).

The Halloween event has the largest SEP event of solar cycle 23, lasting ~2 weeks before returning to the background level (Mewaldt et al., 2005; Jiggens et al., 2014). The largest SEP event associated with ~2500 km/s CME and X17.2 class flare occurred close to the disk center at ~1100 UT on October 28, 2003. Rohen et al. (2005) reported ozone destruction due to the SEP events. Interestingly, daytime aurora was observed at 630 nm over Boston on October 30, 2003 (Pallamraju and Chakrabarti, 2005). The record largest solar flare occurred on November 4, 2003 (~X40 flare; Brodrick et al., 2005). It is also notable that the magnetic storm on November 20, 2003 was the largest magnetic storm in solar cycle 23 (Kataoka et al., 2005b) and caused by the same active region of the Halloween events after one solar rotation.

3.5.11 January 20, 2005: The second largest GLE event

The record high CME speed of 3675 km/s was observed on January 20, 2005 (Gopalswamy 2012). This is the second largest GLE event, with 4200% increase of neutron count rate (Bieber et al., 2013). Geosynchronous bow shock crossing was also observed (Dmitriev et al., 2014). As the industry-specific impacts from this event, National Research Council (2008) noted: "The diversion of 26 United Airlines flights to non-polar or less-than-optimum polar routes during several days of disturbed space weather in January 2005. The flights were diverted to avoid the risk of HF radio blackouts during PCA events. The increased flight time and extra landings and takeoffs required by such route changes increase fuel consumption and raise cost, while the delays disrupt connections to other flights."

3.5.12 July 23, 2012: The STEREO event

The solar maximum of solar cycle 24 (2009–2020) is known as "mini-max" (Zhang et al., 2021). Even in the relatively weak cycle, we witnessed the extreme events. STEREO-A spacecraft, located away from the Earth, observed the record high solar wind speed of 2250 km/s and the magnetic field strength of 109 nT on July 23, 2012 (Russell et al., 2013). The transit time from solar flare to the STERO-A (0.96 AU) was only 14.4 hours. The initial CME speed was ~3400 km/s and the 10 MeV proton flux increased over five orders of magnitude at STEREO-A (Jiggens et al., 2014). Baker et al. (2013a) discussed that this CME could cause another Carrington event, and possibly catastrophic technological impact, if it hit the Earth. Riley et al. (2018) presented an excellent case study of the July 23, 2012 event to discuss the so-called "perfect" storm in detail.

3.5.13 September 10, 2017: The Mars event

Another extreme event occurred in the declining phase of the mini-max solar cycle 24. At the western limb of the Sun, X8.2 flare and a high-speed CME were observed from the Earth on September 10, 2017. The space weather impacts were observed at not only the Earth, but also Mars (Lee et al., 2018; Guo et al., 2018; Jiggens et al., 2019). During the September 2017 event, the same SEP events were detected at the surface of two planets for the first time: the event is registered as GLE72 on the Earth (Kataoka et al., 2018), while it is the largest GLE seen by MSL/RAD on Mars (Guo et al., 2018; Ehresmann et al., 2018; Hassler et al., 2018). The radiation dose was 0.1 mGy at ISS, while it was 0.4 mGy at Mars surface (Berger, 2018). Diffuse aurora was observed during the SEP event (Schneider et al., 2018; Nakamura et al., 2022).

CHAPTER 4

Forecasting space hazards

4.1 Forecasting solar flares

Sunspots have been monitored for 400 years by daily sketch. Flare activity has been continuously monitored 24 hours daily for 7 days a week by Geostationary Operational Environment Satellite (GOES) from 1975. Beautiful coronal images obtained through space telescopes are useful to identify the flaring sites, while the detailed observation of surface magnetic field around the active regions is essential to drive the flare models.

Deep Flare Net is an operational model based on a deep learning technique to forecast the probability of flare occurrence in following 24 hours (https://defn.nict.go.jp/). Using the real-time data of Solar Dynamic Observatory, many features are calculated around sunspots as the input parameters to obtain the probability of M-class flares (Nishizuka et al., 2018). Since the number of X-class flares and larger flares are not sufficient for deep learning, the physics-based model is necessary for forecasting X-class flares and extreme events.

Coronal loops can be modeled as the so-called force-free magnetic field (Wiegelmann and Sakurai, 2012).

$$(\nabla \times \mathbf{B}) \times \mathbf{B} = 0 \tag{4.1}$$

The solution is given by:

$$\nabla \times \mathbf{B} = \alpha \mathbf{B} \tag{4.2}$$

If $\alpha = 0$, the current-free magnetic field is called *potential field*. If α is not constant, it is called *nonlinear force-free field*. The data of vertical magnetic field at solar surface can give the potential field, while horizontal magnetic field is necessary to estimate the nonlinear force-free field. The nonpotential magnetic energy in the active region is sufficient to power the CMEs, the flare-accelerated particles, and the hot thermal plasma in the postflare loops (Emslie et al., 2012). In fact, the latest model suggests that the excess energy of nonlinear force-free field from the potential field gives the upper limit of flare energy (Kusano et al., 2020).

Extreme Space Weather
DOI: https://doi.org/10.1016/B978-0-12-822537-0.00001-6

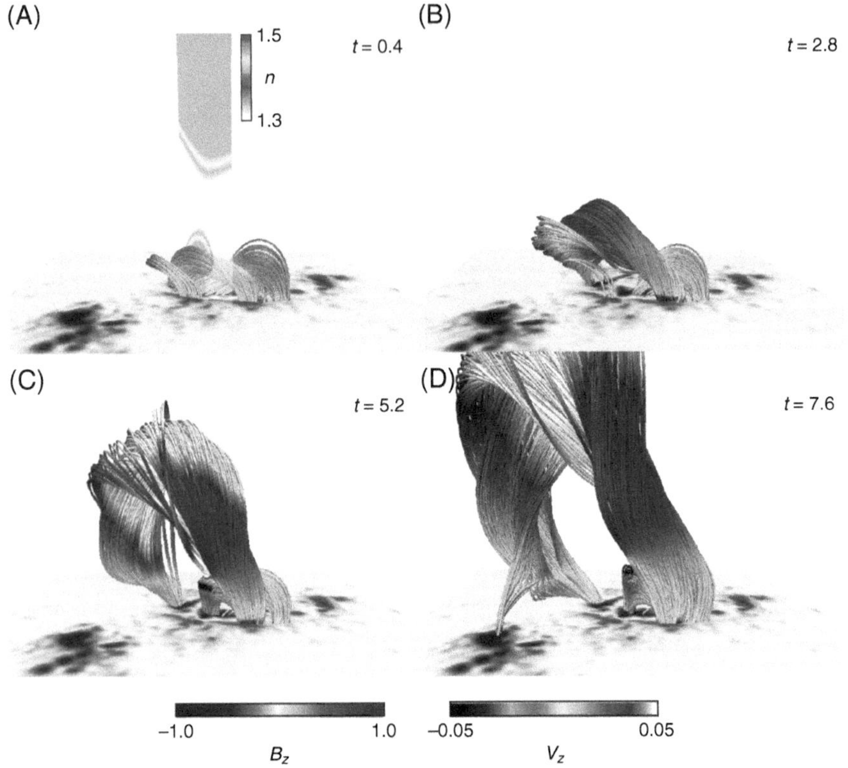

Fig. 4.1 ***Magnetohydrodynamic (MHD) simulation results of eruptive flux tube starting from force-free magnetic field (Inoue et al., 2018).***

Instabilities and loss-of-equilibrium can rapidly deform the force-free field. Fig. 4.1A shows a coronal magnetic field structure above the sunspots as reconstructed using a method to estimate the nonlinear force-free field from the observed vector magnetic field data at the solar surface (Inoue et al., 2018). Starting from the force-free solution, the MHD allows the highly distorted magnetic fields to reconnect to form a single large flux rope (Fig. 4.1B), which rapidly expands outward (Figs. 4.1C and 4.1D). Toriumi and Wang (2019) reviewed many other data-driven simulations in detail.

If the magnetic field configuration meets some criteria, the active region can be explosive to cause large solar flares. Schrijver (2007) suggested that the integrated magnetic flux around the active region is useful to empirically evaluate the possibility to cause flares, which is called R scheme. The physic-based prediction scheme, also known as κ scheme, has been suggested based on their findings that small-scale reconnection induces a macroscopic

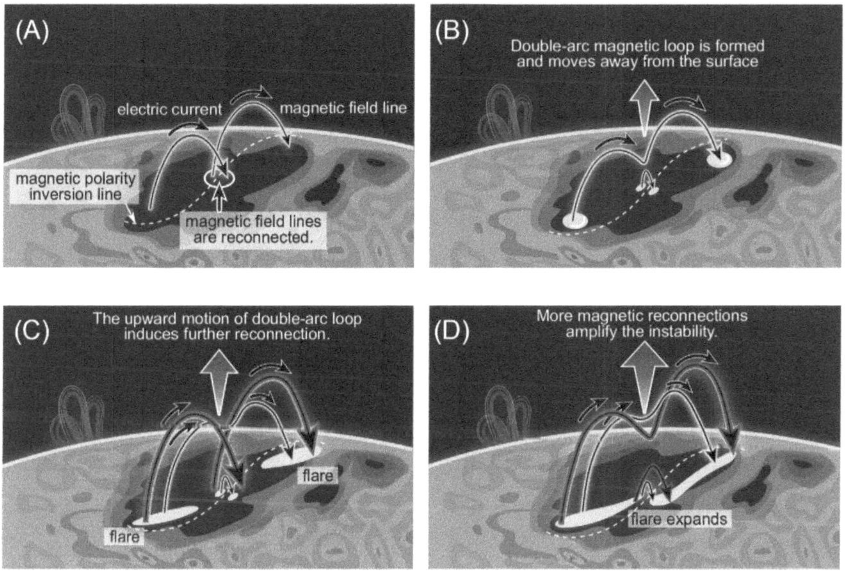

Fig. 4.2 ***The process of solar flare production in the physics-based prediction method of Kusano et al. (2020).*** (A) Electric currents flow along magnetic field lines across the magnetic polarity inversion line on the solar surface, where the magnetic field changes its polarity. (B) Magnetic field lines are reconnected and form a double-arc loop that moves away from the surface due to MHD instability. (C) The upward motion of the double-arc loop induces further magnetic reconnection. A solar flare begins to burst out from the base points of the reconnected field lines. (D) More magnetic reconnections amplify the instability and the solar flare expands (https://en.nagoya-u.ac.jp/research/activities/news/2020/08/research-team-develops-the-first-physics-based-method-for-predicting-large-solar-flares.html). *Copyright: Institute for Space-Earth Environmental Research, Nagoya University.*

double-arc instability, which causes X-class flares (Fig. 4.2). X-class flares are predictable by κ scheme if we have good vector magnetic field observations real-time (Kusano et al., 2020). The next step to earn forecasting lead time would be to predict the detailed evolution of sunspots.

Another challenge is to evaluate whether the predicted flare activity is eruptive or not. Whether the flux rope can erupt as a CME depends on the surrounding large-scale coronal magnetic field because the surrounding coronal field can sometimes overwhelm to confine the flux rope. The future pathway is to investigate this problem by MHD simulation in detail by solving both the global-scale evolution and the small-scale evolution around the active region simultaneously.

4.2 Arrival of coronal mass ejections

SOHO spacecraft at the Lagrange 1 (L1) point (~234 R_E upstream from the Earth) has observed CMEs since 1996 using coronagraphs. The fields-of-view of the coronagraphs are 5 and 30 solar radii (Rs). Twin STEREO satellites orbiting at ~1 AU have also observed CMEs from different angles since 2006. The CME direction and speed can therefore be estimated real-time both manually and automatically, based on such real-time monitoring data. Future coronagraph observations are prepared at both L1 and GEO under the Space Weather Follow-On program, to provide operational resiliency.

It is important to know exactly when the CMEs arrive at or glance off the Earth. The lead time contributes to mitigate the possible magnetic storm-associated hazards. The ENLIL model (Odstrcil and Pizzo, 1999) is the pioneering work to achieve this, utilizing the physics-based data-driven MHD simulation. ENLIL is always running to calculate the background solar wind based on the synoptic solar magnetograms and is ready to calculate the CME (pressure pulse) propagation the moment the forecaster notices. Such a setting is necessary partially because the arrival time depends on slow wind and CIR distribution, which slows down CMEs. Moreover, the CMEs and shocks widely spread or deflected, depending on the background solar wind structure. The shock arrival time can therefore be predicted as an MHD solution from ENLIL.

The latest benchmark studies show that the possible errors are 9 to 11 hours on average with a tendency to early prediction error of −4.0 hours (Wold et al., 2018), which is still comparable with the empirical prediction error of ~11 hours (Gopalswamy et al., 2001). Before the golden age of global MHD simulations, 3D Hakamada-Akasofu-Fry model (Dryer et al., 2004; Fry et al., 2001; Hakamada and Akasofu, 1982) was one of the most successful data-driven operational models for the similar purpose.

The CMEs propagating from the Sun to the Earth can be observed through ground-based measurements. The radio measurement called *interplanetary scintillation* (IPS) is useful to measure CME speed from the appearance of the strong scintillation of radio stars at many different angles. It has been suggested that the arrival-time forecast can be improved (Iwai et al., 2021). Another useful method to indirectly measure the propagating CMEs is global muon detector network (GMDN) observation. Several tens of GeV protons of the GCR are the dominant source of muons in the air shower. The GCR is transiently modified by the CME-associated localized strong magnetic field, and this Forbush-decrease modification can be detected by GMDN. The pioneering work for such application was given by Kuwabara et al. (2006). More recently, Kihara et al. (2021) successfully reproduced a peculiar slow CME event using a sophisticated analysis of the Forbush decrease event using GMDN observations. The CME or shock arrival times at Mars can also be confirmed by Forbush decrease events. The prediction errors of the Mars arrival times as estimated from different methods using STEREO Heliospheric Imagers are ranging from -0.5 to +0.2 days with the standard deviation of ~0.8 days (Freiherr von Forstner et al., 2019).

The SBZ must be predicted to forecast the magnetic storms in addition to CME arrival times. It is, however, still difficult to measure the magnetic field of CMEs in advance, until it is measured in-situ at L1 spacecraft, which gives only 1-hour lead time. To increase the lead time, and to achieve the SBZ forecast, Kataoka et al. (2009) proposed a method to launch a spheromak-type magnetic field as a CME, instead of pressure pulse CME, in the MHD solar wind simulation. This method was then improved by Shiota and Kataoka (2016) to include multiple CMEs simultaneously. This model is called SUSANOO-CME, which enabled us to predict SBZ during the Halloween storms (Fig. 4.3).

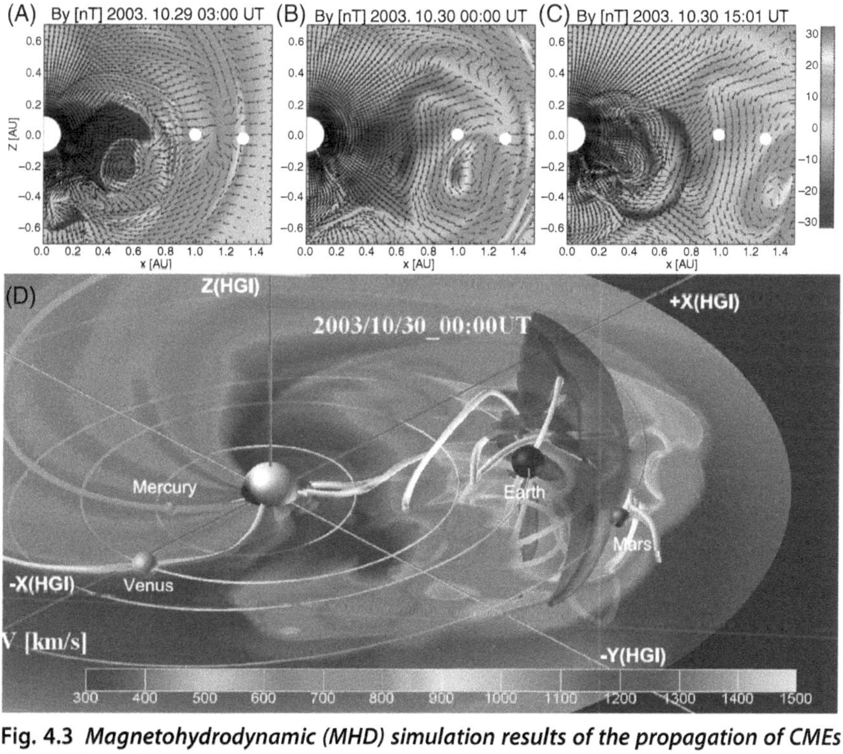

Fig. 4.3 ***Magnetohydrodynamic (MHD) simulation results of the propagation of CMEs during the Halloween event in 2003 (Shiota and Kataoka, 2016).***

The basic set of ideal MHD equations for the solar wind simulations are often written in the normalized conservative forms of Eqs. (1.1)−(1.6) as follows (see Miyoshi and Kusano, 2005):

$$\frac{\partial \rho}{\partial t} + \nabla \cdot (\rho \mathbf{v}) = 0 \tag{4.3}$$

$$\frac{\partial (\rho \mathbf{v})}{\partial t} + \nabla \cdot \left[\rho \mathbf{v}\mathbf{v} + \left(p + \frac{B^2}{2} \right) \mathbf{I} - \mathbf{B}\mathbf{B} \right] = 0 \tag{4.4}$$

$$\frac{\partial \mathbf{B}}{\partial t} + \nabla \times \mathbf{E} = 0 \tag{4.5}$$

$$\frac{\partial e}{\partial t} + \nabla \cdot \left[\left(\frac{\rho v^2}{2} + \frac{\gamma p}{\gamma - 1} \right) \mathbf{v} + \mathbf{E} \times \mathbf{B} \right] = 0 \tag{4.6}$$

where $\mathbf{E}$ is the electric field and $\mathbf{B}$ is the magnetic field, both of which obey the Maxwell's law. The other quantities involved are mass density ρ, velocity $\mathbf{v}$, and gas pressure p. The Maxwell equations can be simplified as

$$\mathbf{E} = -\mathbf{v} \times \mathbf{B} \tag{4.7}$$

$$\nabla \cdot \mathbf{B} = 0 \tag{4.8}$$

The ideal gas is also assumed to close the energy conservation equation as follows.

$$p = (\gamma - 1) \left(e - \frac{\rho v^2}{2} - \frac{B^2}{2} \right) \tag{4.9}$$

Pomoell and Poedts (2018) recently developed a similar MHD solar wind model in Europe, called European Heliospheric Forecasting Information Asset (EUHFORIA), for their operational purpose. Asvestari et al. (2021) showed successful examples of spheromak-type CME simulations by EUHFORIA. Kilpua et al. (2019) summarized the challenges and outlook predicting magnetic structure of CMEs.

The inner boundary model limits the ability of these approaches of ENLIL, SUSANOO-CME, or EUFFORIA models. The background solar wind speed applied at the inner boundary is typically based on the Wang-Sheeley-Arge (WSA) model (Wang and Sheeley Jr., 1990; Arge and Pizzo, 2000; Arge et al., 2004), which empirically derives the solar wind speed from the map of photospheric magnetic field. The metrics of several different models of the inner boundary including WSA model is evaluated (Reiss et al., 2019, 2020).

As a next step, evolution of CMEs in corona (inside of the inner boundary of ENLIL, SUSANOO-CME, or EUHFORIA) must be self-consistently solved. The data-driven MHD-based global corona and solar wind models (Nakamizo et al., 2009; van der Holst et al., 2014) are the straightforward pathway to improve the operational forecasting simulations. Based on such realistic global corona and solar wind, the dynamic evolution of model CMEs can be solved from corona to the solar wind (Jin et al., 2017). Furthermore, although it takes time to be operational use in future, Torok et al. (2018) presented cutting-edge Sun-to-Earth modeling even from the initiation of CMEs for the Bastille Day event on 15 July, 2000. The resultant flux rope is similar in character to the observed magnetic cloud at the Earth, but it arrives ~8.5 hours later and ~15° N from the Earth position, with field strengths that are weak by a factor of ~1.6.

During the solar cycle 25, new successful missions such as Parker Solar Probe, Solar Orbiter, and Bepi-Columbo are exploring the inner heliosphere inside the 1 AU orbit. These missions provide additional multipoint in-situ observation data of the upstream solar wind, which will contribute to improve the existing physics-based models and their accuracy to predict the arrival time, SBZ, and amplitude of possible space weather influence at the Earth, the Moon, and the Mars.

4.3 Forecasting substorms

In-situ monitoring of the solar wind parameters, such as flow speed, density, temperature, and vector magnetic field, at L1 point (234 R_E upstream) by ACE (since 1998) or DSCOVR (since 2018) is used to forecast the space weather in the magnetosphere and ionosphere. Although the lead time is only ~1 hour, the accuracy is significantly better than the predicted values of the solar wind or CME arrival models with a lead time of a few days. The magnetosphere–ionosphere system is sensitive to the SBZ, and reliable SBZ forecast with a few days lead time has been a challenge. Inputting the solar wind parameters, the empirical models such as Tsyganenko models (Tsyganenko, 1989; Tsyganenko and Stern, 1996; Tsyganenko and Sitnov, 2005) and Weimer models (Weimer, 1995, 2005) provide the averaged magnetosphere configuration and averaged ionospheric convection pattern, respectively. The resultant magnetospheric and ionospheric responses are monitored by GOES at GEO, many ground-based instruments including magnetometer networks such as SuperMAG, and HF radar network such as SuperDARN.

The energy flow from the solar wind into the magnetosphere can be roughly formulated by Poynting flux, which is proportional to VB^2 (the solar wind speed V multiplied by the motional E field = VB):

$$\varepsilon_{Akasofu} = VB^2 f l^2, \tag{4.10}$$

where l is assumed to be 7 R_E, and the efficiency f is defined as:

$$f \equiv \sin^4\left(\frac{\theta}{2}\right), \theta = \begin{cases} \tan^{-1}\left(B_Y / B_Z\right) \text{ if } B_Z > 0 \\ \pi\text{-}\tan^{-1}\left(B_Y / B_Z\right) \text{ if } B_Z < 0 \end{cases} \tag{4.11}$$

This Poynting-flux expression is called *Akasofu's epsilon* (Akasofu, 1981), which is useful to predict the amplitude of substorms, although the physics behind is complex (Ebihara et al., 2019).

Global MHD simulation is a powerful tool for space weather forecasting because they self-consistently reproduce the global ionospheric patterns of plasma convection, field-aligned currents, as well as the global magnetospheric configuration. It has been confirmed that the basic steady-state patterns of field-aligned current (Iijia and Potemra, 1978) and auroral oval are reasonably reproduced (Tanaka, 2007).

In the global MHD simulations, the solar wind is given at the outer boundary, and at the inner boundary of a few R_E, the ionosphere is assumed to be the thin conductive shell with the conductivity tensor $\tilde{\sigma}$ that closes the field-aligned current $J_{||}$. Modifying the Eq. (1.38), this relationship becomes:

$$\nabla \cdot \left(\tilde{\sigma} \nabla \Phi\right) = J_{||}, \tag{4.12}$$

where Φ is the ionospheric potential. The actual magnetosphere–ionosphere coupled region between a few R_E and the ionosphere was therefore omitted. The potential drop in the magnetosphere–ionosphere coupled region must be self-consistently treated in future, but in the current models, this is approximated by the simplified relation that the ionospheric conductivity increases at regions in the presence of upward field-aligned current (Tanaka, 2000). The basic set of equations for the magnetospheric part is often written in the conservative forms of Eq. (4.3)−(4.9).

Recently, the dynamic variations of magnetosphere and ionosphere, such as substorms and SCs, can also be reproduced by the global MHD simulations (Fujita et al., 2003a, 2003b; Kataoka et al., 2004; Ebihara and Tanaka, 2015). One of the most outstanding results is the detailed substorm reproduction (Fig. 4.4). The global simulation can therefore forecast the

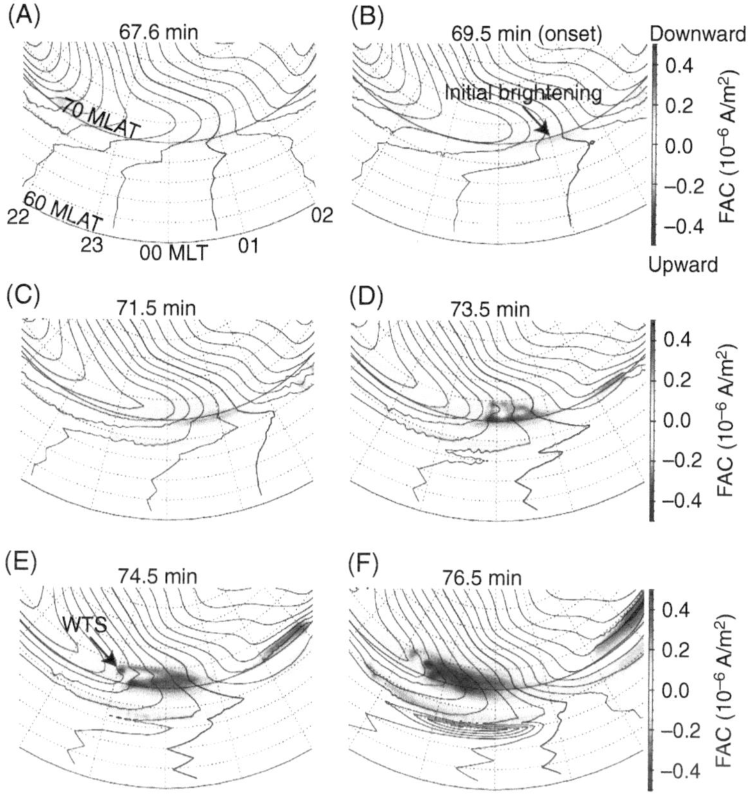

Fig. 4.4 ***Magnetohydrodynamic (MHD) simulation results of the field-aligned current (FAC) and convection patterns during the substorm expansion (Ebihara and Tanaka, 2015).***

resultant GMD as observed by ground-based magnetometers. The real-time global MHD simulation called REPPU (Tanaka, 2015) is used for space weather operational forecast at NICT, Japan.

Extreme simulations have been performed to reproduce the Carrington-class magnetic storms by maximizing the solar wind parameters (Li et al., 2005; Ngwira et al., 2014). The extreme cases such as super SCs have also been studied using the global MHD simulations (Fig. 4.5) to make a physics-based estimate of the super SC amplitude (Kubota et al., 2015; Welling et al., 2021). The spiky waveforms of anomalous SCs (Araki, 2014) have been reproduced by Kubota et al. (2015). As a theoretical discussion, Tsurutani and Lakhina (2013) estimated the possible parameters of the perfect interplanetary shock that can cause extreme SCs. Super substorms of AL index ~2500 nT have also been simulated by Ebihara et al. (2019).

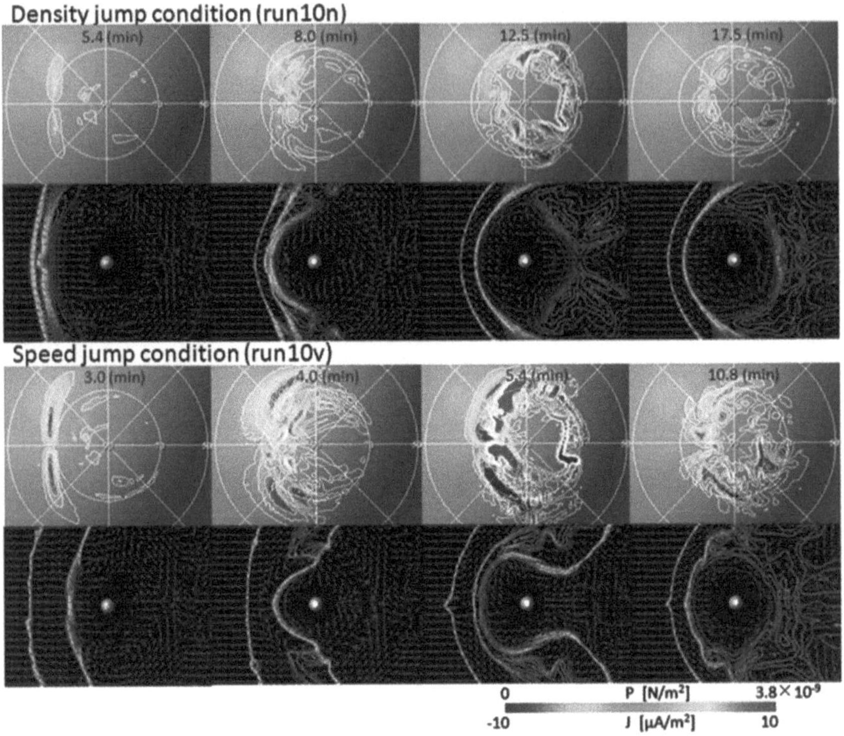

Fig. 4.5 ***Magnetohydrodynamic (MHD) simulation results of the arrival of extreme sudden commencements (SCs) for density jump condition (run10n) and speed jump condition (run10v) with a tenfold increase in the solar wind dynamic pressure.*** The upper panels of each case show the field-aligned current (FAC) in the polar ionosphere. The lower panels show the pressure and velocity vector unit at the equatorial plane in the magnetosphere. The sunward direction is to the left (Kubota et al., 2015).

Tsurutani et al. (2015) discussed that density jump in the solar wind likely triggers the observed super substorms.

4.4 Forecasting magnetic storms

Magnetic storms have been monitored by the Dst index since 1957. Burton et al. (1975) proposed the simplest expression to describe the time variation of the Dst index as follows:

$$\frac{dD_{st}}{dt} = Q - \frac{1}{\tau} D_{st}, \tag{4.13}$$

where τ is the decay scale of typically 8 hours. The slow recovery is well recognized as a result of charge-exchange collisions between ring current protons and low-energy neutrals called the geocorona. There are other processes that could contribute to the recovery, such as (a) the charge-exchange loss of singly charged oxygen ions (O+), (b) flow-out of ring current ions across the dayside magnetopause, and (c) pitch-angle scattering of ring current ions due to field-line curvature of near-Earth magnetotail or resonant interactions with electromagnetic ion cyclotron (EMIC) waves and consequent precipitation into the atmosphere. The energy input Q is defined as:

$$Q = \begin{cases} a(VB_S - E_c), VB_S > E_c \\ 0, VB_S < E_c \end{cases} \tag{4.14}$$

The energy input proportional to the solar wind speed V only works during the southward magnetic field B_S. O'Brien and McPherron (2000) empirically determined the parameters a and E_c to be −4.4 and 0.49 mV/m, respectively. Keika et al. (2015) reproduced the Carrington event by adding the flow-out loss term as a function of solar wind density. Such empirical tunings work well for forecasting the Dst index using the real-time solar wind parameters obtained at L1 point or the predicted values from the solar wind and CME simulations.

During intense magnetic storms, the strong ring current modifies the ambient magnetic field structures of the magnetosphere. Such effects are included in the recent empirical model of the magnetospheric magnetic field, called TS05 model (Tsyganenko and Sitnov, 2005). Also, the magnetosphere can be extremely compressed during the occurrence of great magnetic storms. The empirical model of the magnetopause location, including such an extreme situation, was also developed (Shue et al., 1998) and frequently used in space weather community.

The Dst index only represents the 1-hour averaged total accumulated energy in the inner magnetosphere. In reality, substorms play an essential role to inject charged particles in the inner magnetosphere in a short time scale. As a result, the ground-based magnetometers respond differently at different local time. To predict the super magnetic storms and their localized effects, which is important to mitigate the possible GIC hazards, physics-based models must be introduced.

The first version of the operational GIC model was released in 2017, using 1D conductivity model (Fernberg, 2012) and is available on NOAA/SWPC website. A different version incorporating the full 3D effects of Earth conductivity structures are also available. In Japan, magnetotelluric observations have been continued since 1958 (Fujii et al., 2015). However,

Japan Meteorological Office decided to stop the Magnetotelluric observations at Memambetsu, Kakioka, and Kanoya in 2020. Instead, using a modern system, GIC is monitored in Japan in Kanto areas (Watari et al., 2021). One of the best physics-based GIC models is a time-domain simulation, which successfully reproduced the storm-time GICs in Japan (Nakamura et al., 2018).

Extreme situations occur also in the subauroral latitudes, which include the plasmasphere and low-density *trough* in the ionosphere. *Subauroral polarization stream* (SAPS, Foster and Burke, 2002; Foster and Vo, 2002) or *subauroral ion drift* (Fig. 4.6), as identified by the enhanced ionospheric plasma flow, are interesting phenomena appeared during magnetic storms to understand the interplay between the ionosphere and magnetosphere via field-aligned currents (Goldstein et al., 2005). *Stable auroral red* (SAR) arcs also coexist in the subauroral region (Foster et al., 1994). The role of asymmetric distribution of the ring current, called *partial ring current*, and the connection to the ionospheric current system to cause SAPS are discussed by coupled simulations (Kataoka et al., 2007, 2009; Ebihara et al., 2009).

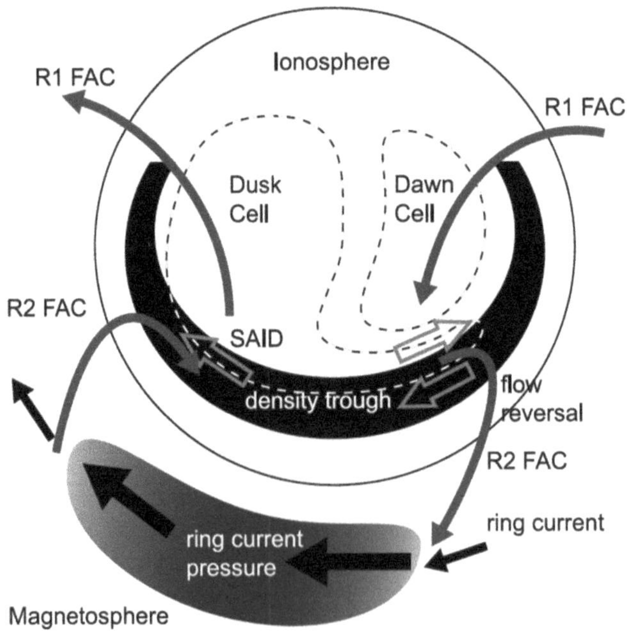

Fig. 4.6 ***Schematic illustration of subauroral ionospheric phenomena and the partial ring current in the magnetosphere (Kataoka et al., 2007).***

The Boltzmann equation is the governing equation for ring current plasma. The guiding-center approximation leads to the drift kinetic equation (e.g., Amano et al., 2011), as described by the time evolution of the five-dimensional phase-space density $f(\mathbf{x}, v_{\parallel}, \mu)$ as a function of position $\mathbf{x}$, parallel velocity $v_{\parallel}$, and the first adiabatic invariant $\propto = mv_{\perp}^2 / 2B$.

$$\frac{\partial f}{\partial t} + \mathbf{v} \cdot \nabla f + \frac{\partial v_{\parallel}}{\partial t} \frac{\partial f}{\partial v_{\parallel}} = 0, \tag{4.15}$$

where the particle velocity and the acceleration follow the conventional particle drift equations. The time development of the electromagnetic fields is governed by the Maxwell equations as follows:

$$\frac{\partial \mathbf{B}}{\partial t} = -\nabla \times \mathbf{E} \tag{4.16}$$

$$\nabla \times \mathbf{B} = \mu_0 \mathbf{J} \tag{4.17}$$

where displacement current is ignored as it is not important for the low-frequency phenomena.

Comprehensive Ring Current Model (CRCM) is the successful model (Fok et al., 2001), which was recently improved as Comprehensive Inner-Magnetosphere Ionosphere (CIMI) model (Fok et al., 2021). Averaging the third term of Eq. (4.15), the bounce-averaged four-dimensional Boltzmann equations were solved in CRCM or CIMI models. In large magnetic storms, O+ becomes the dominant ion species. During extreme events, like the Halloween storm in October 2003, the O+/H+ energy density ratio is as high as 10–20 (Nose et al., 2005). The phase-space density must therefore be prepared for not only protons but also oxygen ions.

Extreme magnetic storms have also been discussed using CRCM (Kataoka et al., 2005b; Ebihara et al., 2005). Fig. 4.7 shows the quantitative comparison between the observation data and the modeling results from the CRCM simulation for the largest magnetic storm during the solar cycle 23.

Baker et al. (2013a) discussed that the CME on 23 July, 2012, with 2000–3000 km/s initial speed, could lead to another Carrington-class magnetic storm with the Dst index of −1182 nT if it hit the Earth, assuming the worst case SBZ of ~70 nT. Fortunately, the CME was directed to the STEREO-A spacecraft, which was 125° away from the Earth. Ngwira et al. (2013a) also simulated the magnetospheric response during the super CME without artificially changing the SBZ and obtained severe GMD comparable with those of the Halloween 2003 storms.

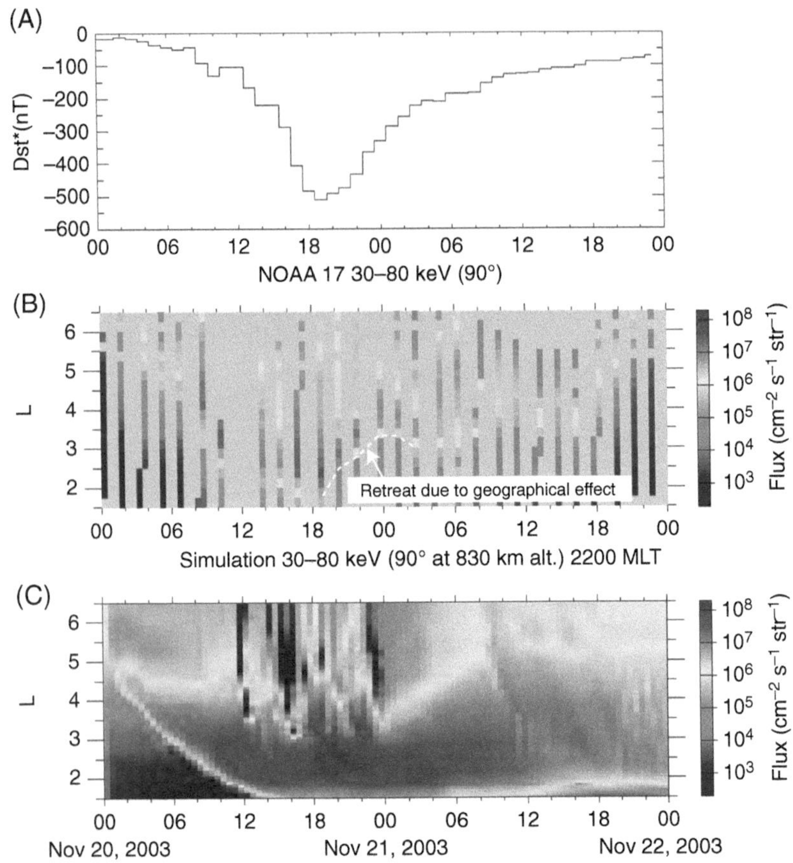

Fig. 4.7 ***Comparison between observation data and simulation data for the evolution of 30–80 keV proton flux during the largest magnetic storm of solar cycle 23.*** Top panel shows the Dst index corrected by the solar wind dynamic pressure (Ebihara et al., 2005).

4.5 Forecasting radiation belts

Since 1975, GOES has been monitoring MeV electron flux at GEO at the edge of the outer radiation belt, while POES has been monitoring the sub-MeV electrons at LEO crossing the subauroral zones by the polar orbit. However, it has been a challenging task to predict the dynamic variation of the outer belt electrons using fully physics-based models because there are many important modules (Fig. 4.8) that are relevant to the loss and source mechanims of these electrons.

Empirical and probability forecast is useful for such a complex target. As an example, an empirical forecast of outer belt electrons has been operated at NICT, Japan (Sakaguchi et al., 2013; 2015). Imagine how the

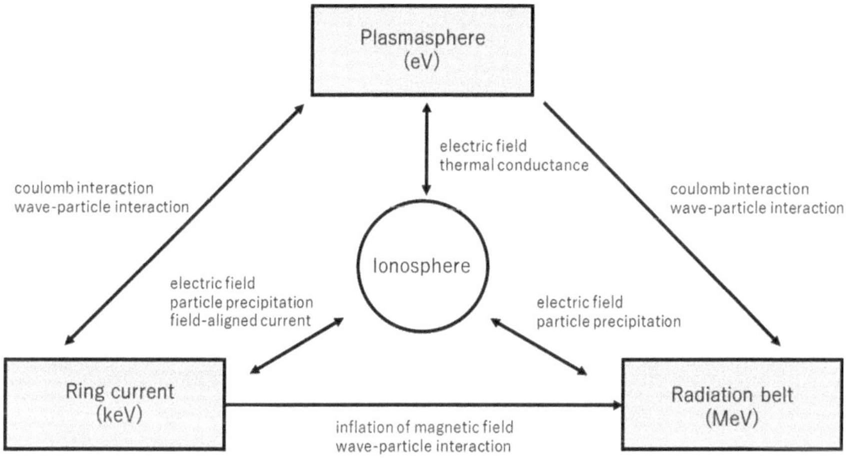

Fig. 4.8 ***Cross-energy and cross-region couplings between the inner magnetosphere and the ionosphere.*** See also Ebihara (2019).

probability weather forecast of rainy days works during both normal and stormy weathers. During normal weather, robust forecast is possible by knowing the arrival time of a cold front, while during stormy weather, the robust forecast is possible by knowing the arrival time of a typhoon. As shown in Fig. 4.9, the similar idea is applicable to space weather forecast as well. Knowing the arrival of *interplanetary shocks* (CME marker) and *stream interfaces* (CIR/SIR maker), we can empirically predict the probability of the large enhancement of outer belt electrons for a few days after their arrival (Kataoka and Miyoshi, 2006).

The basic parameter to describe the radiation belts is the differential flux of electrons at certain kinetic energy (or momentum p) at certain position as denoted by L value (see Section 2.5). As the starting point of a physics-based forecast model, phase-space density f and the diffusion quantitatively can deal with a collection of energetic particles in a straightforward way. The Fokker–Planck equation to model the radiation belts (Schultz and Lanzerotti, 1974) in a simplified form is:

$$\frac{\partial f}{\partial t} = L^2 \frac{\partial}{\partial L}\left(\frac{1}{L^2} D_{LL} \frac{\partial f}{\partial L}\right) + \frac{1}{p^2}\frac{\partial}{\partial p}\left(p^2 D_{pp} \frac{\partial f}{\partial p}\right) + Sources - Losses, \quad (4.18)$$

where f is the distribution function of energetic particles, L is L-value, and p is momentum. The first term indicates radial diffusion and the second term indicates energy diffusion. Many other additional terms such as pitch-angle diffusion and associated atmospheric losses can also be introduced in

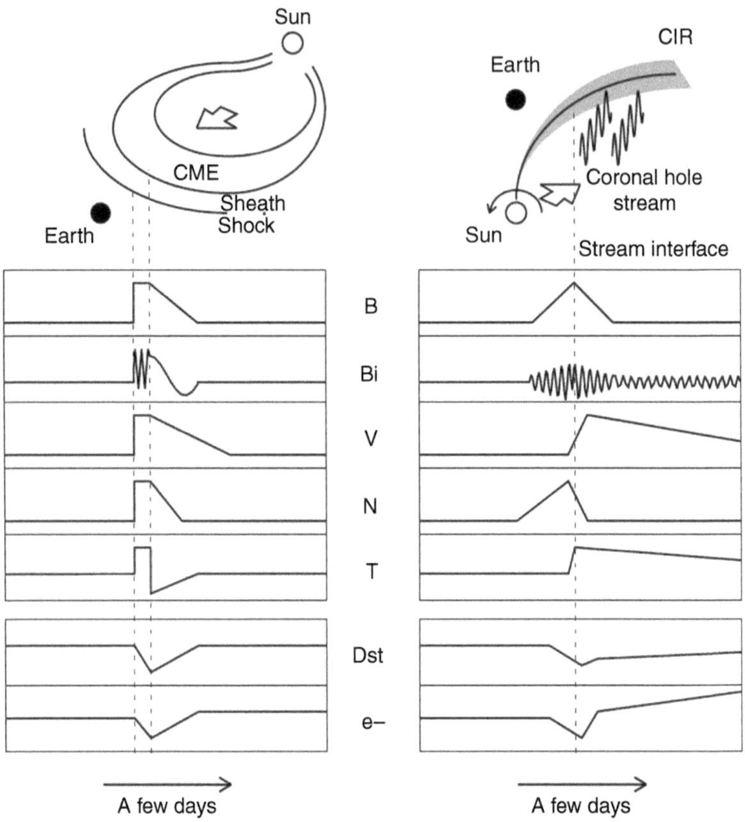

Fig. 4.9 ***Schematic picture for empirical space weather forecast of outer belt electrons after the arrival of interplanetary shock and stream interface.*** Top panels show the typical variation of solar wind parameters: magnetic field, speed, density, and temperature. Bottom two panels show the typical variation of the Dst index and electron flux at the geosynchronous orbit (Kataoka Miyoshi 2006).

Eq. (4.18). One of the successful physics-based operational models is based on the Fokker-Planck equation, combined with the data assimilation of real-time observation data (Subbotin et al., 2010).

As the sources of the outer belt, it is clear from the phase space density analysis of in-situ observation data that the internal source of nonadiabatic energization occur at outer belt (Reeves et al., 2013; Miyoshi et al., 2021), and whistler-mode chorus waves can account for it (Thorne et al., 2013). The cross-energy coupling process via the whistler mode waves play the important role, and the interplay among different energy populations ranging from the plasmasphere electrons and radiation belt electrons

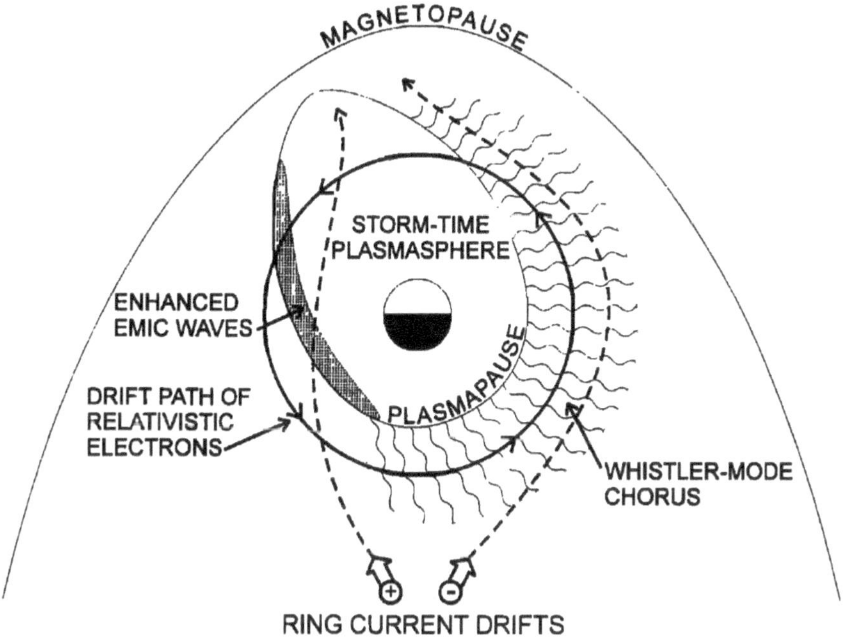

Fig. 4.10 ***Schematic illustration of different types of plasma waves in the magnetosphere which can scatter relativistic electrons (Summers et al., 1998).***

are therefore important (Miyoshi et al., 2013, 2018). The drift-resonance between ULF waves and drift-electrons also plays an important role for the formation of the outer belt (Elkington et al., 1999; Teramoto et al., 2019). These Pc5 ULF waves are generated by the variation of the solar wind dynamic pressure or Kelvin-Helmholtz instability around the magnetopause. The double accelerations worked by ULF waves and chorus waves are also discussed (Katsavrias, et al., 2019). Further, time-domain structure evolved from whistler mode waves (Mozer et al., 2016) and kinetic Alfvén waves may play important roles (Chaston et al., 2017) to disturb the outer belt electrons. The basic plasma waves which can scatter energetic electrons are graphically summarized in Fig. 4.10.

The source term is therefore enhanced by plasma injections into the inner magnetosphere associated with substorms, while the loss term is enhanced during the magnetic storm main phase. The loss of electrons is two-fold (Green and Kivelson, 2004): magnetopause shadowing (Kim and Lee, 2014; Mauk et al., 2016) and atmospheric loss by pitch-angle scattering due to enhanced plasma waves, which is explained in more detail below. It is also interesting to note that the extreme flux enhancement of outer belt electrons was observed during a particular time of a small-loss event, due to magnetosphere inflation (Kataoka and Miyoshi, 2008). The global MHD simulation and magnetic storm simulation as introduced in the previous sections are therefore necessary to be fully developed for the accurate prediction of radiation belts.

The loss processes of radiation belt electrons have been reviewed by Millan and Thorne (2007) and Blum and Breneman (2020). Resonant wave–particle interactions occur when electrons interact with plasma waves whose frequency ω is Doppler shifts to a multiple (n = 0, ±1, ±2) of the relativistic electron gyrofrequency.

$$\omega - k_{\parallel} v_{\parallel} = \frac{n\omega_{ce}}{\gamma} \tag{4.19}$$

where γ is the relativistic Lorentz factor and $k_{\parallel}$ and $v_{\parallel}$ are the components of wave vector and particle velocity, respectively, along the direction of the ambient magnetic field. The resonant wave–particle interaction violates the first two adiabatic invariants, leading to the pitch-angle diffusion.

The diffusion coefficient, as well as source and loss terms of Eq. (4.18), therefore depend on the complex wave–particle interactions in the inner magnetosphere. For example, as the simplest form, Kennel and Petschek (1966) discussed that the electron energy E_R resonant with whistler mode waves with frequency ω is given by:

$$E_R = \frac{B^2}{2\mu_0 N_e} \frac{\omega_{ce}}{\omega} \left(1 - \frac{\omega}{\omega_{ce}}\right)^3 \tag{4.20}$$

The resonant energy depends on the ambient parameters such as magnetic field B, electron density N_e, and the electron cyclotron frequency ω_{ce}

$$\omega_{ce} = \frac{qB}{m_e}, \tag{4.21}$$

where m_e is the electron mass and q is the elementary charge. The whistler mode waves are excited by the free energy of distorted phase-space density of energetic electrons close to the resonant energy. Kurita et al. (2018) showed evidence of rapid acceleration of energetic electrons by whistler mode chorus waves through the phase trapping. Such a nonlinear interaction between plasma waves and charged particles is essential for the dynamic variation of radiation belt.

The related phenomena at the atmosphere and LEO include isolated proton auroras (Miyoshi et al., 2008), microburst (Nakamura et al., 1995), and relativistic electron precipitations (REP) at ISS (Kataoka et al., 2016, 2020a).

These REP events at ISS also cause a radiation dose on astronauts during EVA (Ueno et al., 2019).

The direct evidence of loss-cone precipitation through wave-particle interactions was identified by the Arase satellite (Kasahara et al., 2018). Both EMIC waves and whistler mode chorus waves cause the MeV electron precipitation (Tanaka et al., 2019; Kataoka et al., 2020a). Shoji et al. (2021) discovered the direct generation process of EMIC waves through the formation of proton hill. The chorus waves precipitate wide-energy electrons from 10 keV to more than MeV through the nonlinear wave particle interactions to cause the mesospheric ozone loss (Mironova et al., 2015; Miyoshi et al., 2021).

4.6 Forecasting ground level enhancements

Ground level enhancement (GLE) events are monitored real-time by more than a dozen of neutron monitors operating worldwide. The newest one is Syowa Station operating since 2018 (Kato et al., 2021). GOES also provides >100 MeV proton data every 5 minutes.

These data are used to constrain the physics-based aviation-dose forecast model, called WASAVIES (Kataoka et al., 2014, 2018; Sato et al., 2018b). As an example, for the GLE 72 (Table 2.4), the global dose rate map at flight altitude is shown in Fig. 4.11. It is important to select the right

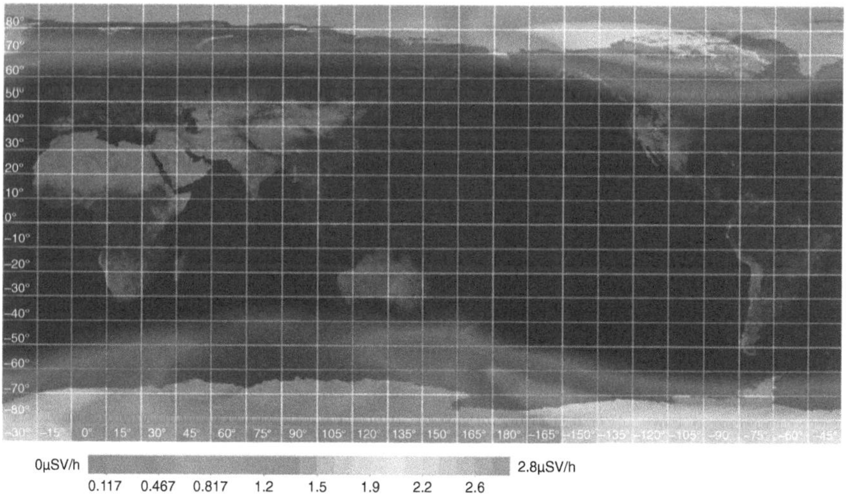

Fig. 4.11 ***Global map of effective dose rate at 12 km altitude at 1800 UT on September 10, 2017 (Kataoka et al., 2018).***

route to avoid the unnecessary radiation dose during a GLE event. For similar purpose, nowcast models of aviation dose such as AVIDOS (Latocha et al., 2009), NAIRAS (Mertens et al., 2009) are also used worldwide. WASAVIES was developed to be operational under the support of Japan's PSTEP project in 2015-2020, to fill the gap between scientific research and operational forecast (Kusano et al., 2021).

The pitch-angle distribution of the solar protons is not isotropic and varies from being beam-like to isotropic. In WASAVIES, the streaming transport and diffusion of energetic protons are solved time-backward starting from the Earth in the 1D coordinate along the Parker spiral, using the *stochastic differential equation* (SDE) method (Kubo et al., 2015), and setting all possible free parameters in advance.

The governing equation of SEP and GLE particles is known as the *focused transport equation*, which is given in the simplest form of negligible convection and adiabatic deceleration (Ruffolo, 1991) by:

$$\frac{\partial f}{\partial t}+\mu v\frac{\partial f}{\partial s}+\frac{1-\mu^2}{2L}v\frac{\partial f}{\partial \mu}-\frac{\partial}{\partial \mu}\left(D_{\mu\mu}\frac{\partial f}{\partial \mu}\right)=Q, \tag{4.22}$$

where $f(s,v,\mu,t)$ is the distribution function on a given magnetic field line, s is the coordinate measured along the magnetic field, v is particle speed, μ is pitch angle cosine of a particle, $L=-B/(\partial B/\partial s)$ is the focusing length determined by the mean magnetic field $B(s)$, and $D_{\mu\mu}$ is the pitch angle scattering coefficient and $Q(s,v,\mu,t)$ is the source function, determined by the processes that accelerate particles in the solar corona and solar wind.

Separately, the proton trajectories are back-traced from the top-of-atmosphere of neutron monitor observatories using the Tsyganenko geomagnetic field model (Tsyganenko and Sitnov, 2005) to identify the corresponding incident differential flux of protons at the Earth orbit. The normalized neutron counts at ground and the radiation dose in the air are also calculated in advance using PHITS (Sato et al., 2018c), an air shower simulation code. After detecting GLEs, WASAVIES is designed to select the best simulation results, which minimizes the difference with the observation data, to make the physics-based forecast real-time.

It is also noteworthy that the results obtained from WASAVIES have been simplified for space weather operations. For example, during the GLE 72 event, complicated time evolution was found in the energy spectra of protons, including the double-peak structure (Bruno et al., 2019), which cannot be explained by a simplified model.

The straightforward way to model SEP and GLE is to solve the acceleration of the solar protons in the solar corona and solve their propagation in the interplanetary space. The arriving solar protons at the top of atmosphere can then also be calculated. Although the computational cost of such forward simulations is high for operation purpose till date, it is expected to decrease in future. Modeling efforts of GCR (background-level of GLE) and SEP events (not limited to GLE events) and their future challenges are discussed in Sections 5.1 and 5.2, although they are also relevant for the Earth.

4.7 Metrics for evaluating different forecast models

As shown in the previous Sections, many different models exist for the same purpose. Emergence of useful machine learning techniques in the last decade especially increased the publication of application papers, and space weather forecast is not an exception (Camporeale, 2019). Liemohn et al. (2021) summarized that common performance metrics are important to evaluate and compare different models. Some of simplest skill scores often used in this field are introduced here.

Cross-correlation efficient has been historically used in many studies to evaluate the similarities of two different data sets. Other metrics include root-mean-square error (RMSE) and prediction efficiency (PE):

$$\mathrm{RMSE} = \sqrt{\frac{1}{N}\sum_{i=1}^{N}(M_i - O_i)^2} \tag{4.23}$$

$$\mathrm{PE} = 1 - \frac{\sum_{i=1}^{N}(M_i - O_i)^2}{\sum_{i=1}^{N}(O_i - \overline{O_i})^2} \tag{4.24}$$

where M_i is the ith modeled data and O_i is the ith observation data. A small RMSE value and PE closer to 1 is preferred.

Other practical metrics have also been introduced. Determining a threshold, we can judge if prediction hit or missed. Using the numbers of hit H and missed M, the probability of detection (POD) can be calculated as:

$$\mathrm{POD} = \frac{H}{H + M} \tag{4.25}$$

Other numbers such as false alarms F and correct negatives N are also important for studying functioning of the prediction models. The probability of false detection (POFD) is therefore defined as:

$$\text{POFD} = \frac{F}{F+N} \tag{4.26}$$

Combining these two metrics, receiver operating characteristic (ROC) curve can be drawn by setting many different thresholds, plotting POFD in the horizontal axis with POD in the vertical axis. The area below the ROC curve determines the high skill of the models.

CHAPTER 5

Toward the Moon and Mars

5.1 Predicting galactic cosmic rays (GCR)

Do you want to travel to Mars during solar maximum or solar minimum? This is a good question to start this Chapter. The first concern is the background dose by GCR. GCR comprises 98% baryons and 2% electrons. The baryonic component comprises 87% protons (hydrogen nuclei), 12% alpha particles (helium nuclei), and 1% heavy nuclei with atomic numbers up to 92 (uranium). Hereafter, we focus on the protons as the dominant player to contribute to the background radiation dose.

GALPROP is known as the best calibrated model to solve the transport and diffusion of GCR in our galaxy (Strong et al., 2007). The heliospheric magnetic field modulates the GCR protons before they are observed at the Earth. The heliospheric magnetic field structure changes along the solar cycles. CREME96 is known as the standard operational model for the SEE rate prediction considering the *solar modulation* (Tylka et al., 1997).

It is impossible to directly measure these GCR protons at the Earth surface, as they hit the atmospheric atoms and molecules to cause air shower. The air shower includes muons and neutrons, which can reach the ground. The muon and neutron count rates at ground are therefore good indicators of the proton flux of primary cosmic rays.

GCR intensity represented by neutron count rate shows 11-year cyclic variation, roughly anticorrelated with sunspots (Fig. 5.1). The modulated proton spectra of the GCR can be modeled by a modulation parameter φ. Assuming the spherical symmetric heliosphere, Gleeson and Axford (1968) derived that the differential intensity J of a cosmic ray nucleus at 1 AU is given as:

$$J(T,\varphi) = J_{\mathrm{LIS}}(T+\Phi)\frac{T(T+2T_r)}{(T+\Phi)(T+\Phi+2T_r)}, \tag{5.1}$$

where $\Phi = (Ze/A)\varphi$, T is the kinetic energy (in MeV per nucleon) of cosmic nuclei with charge number Z and mass number A. The rest proton mass energy T_r = 938 MeV. J_{LIS} denotes the *local interstellar spectrum* (LIS) of cosmic ray nuclei. Usoskin et al. (2005) have reconstructed the modulation parameter for the last 50 years, using worldwide neutron monitor network, while Muscheler et al. (2016) used cosmogenic-radionuclide and sunspot

Extreme Space Weather
DOI: https://doi.org/10.1016/B978-0-128-22537-0.00004-1

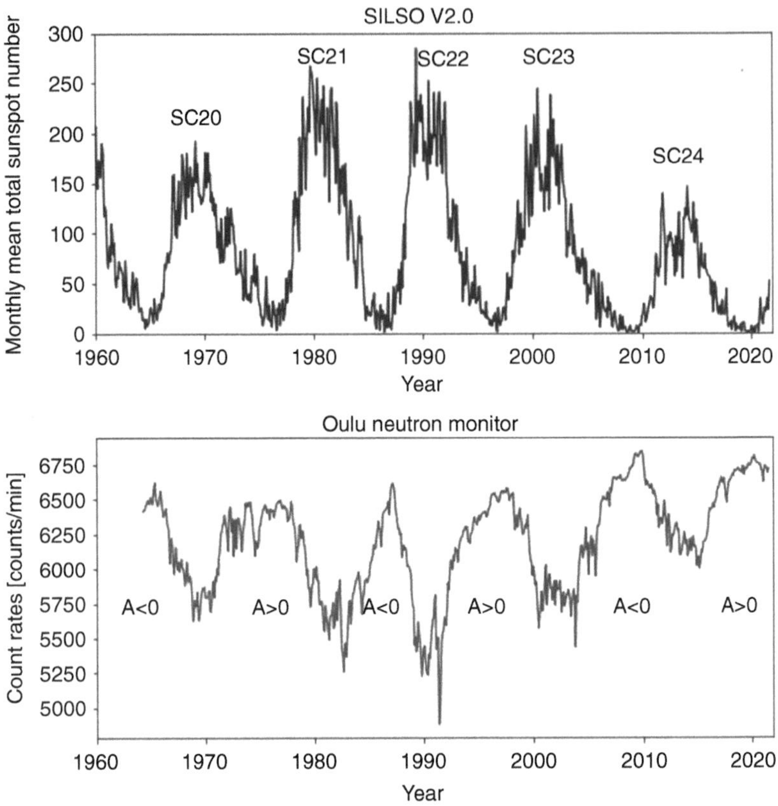

Fig. 5.1 ***Sunspot number variation and neutron counts at Oulu cosmic ray station.***

records to reconstruct the 400-year-long modulation parameter (Fig. 5.2). Matthia et al. (2013) provided a ready-to-use model to reconstruct GCR from sunspot number. Typical GCR spectra during solar maximum and minimum are shown in Fig. 5.3.

Taking a close look at Fig. 5.1, one may notice the asymmetric time profiles of GCR in odd and even solar cycles, that is, flat-top and peaky-top variation. The true cycle length is therefore 22 years, which is associated with the Hale magnetic cycle of the Sun (Fig. 5.4). The question is thus updated as follows: Do you want to travel to Mars during solar maximum or solar minimum, in odd or even solar cycles?

The Parker equation (Parker, 1965) governs the transport and diffusion of GCR in the heliosphere:

$$\frac{\partial f}{\partial t} = \nabla \cdot \left(\tilde{\kappa} \cdot \nabla f \right) + \mathbf{v} \cdot \nabla f - \frac{1}{3} \left(\nabla \cdot \mathbf{v} \right) \frac{\partial f}{\partial \ln p}, \tag{5.2}$$

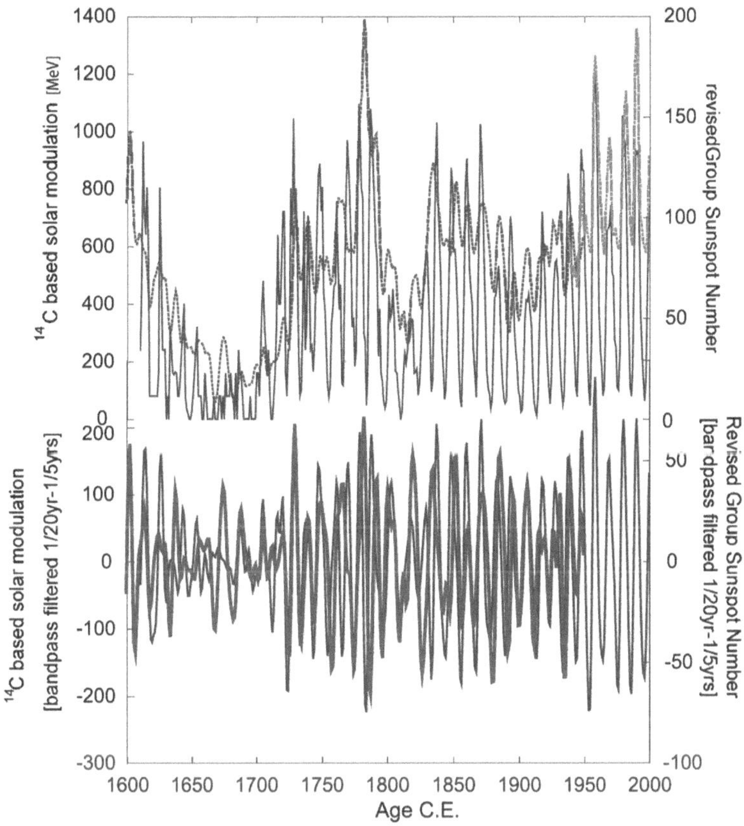

Fig. 5.2 ***Solar modulation parameter reconstructed for 400 years (Muscheler et al., 2016).***

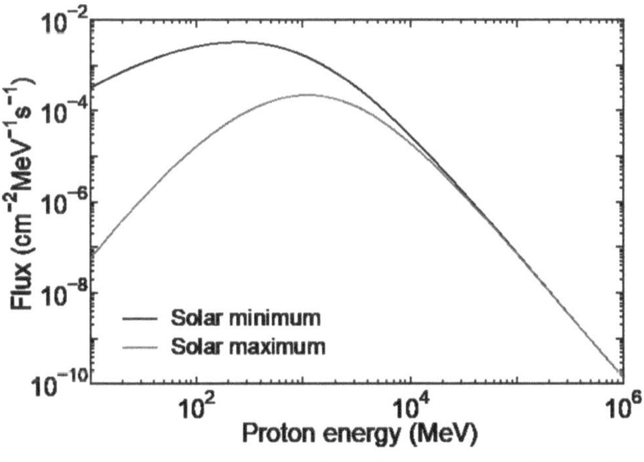

Fig. 5.3 ***GCR energy spectra during the solar minimum and solar maximum.*** *See also Matthia et al. (2013).*

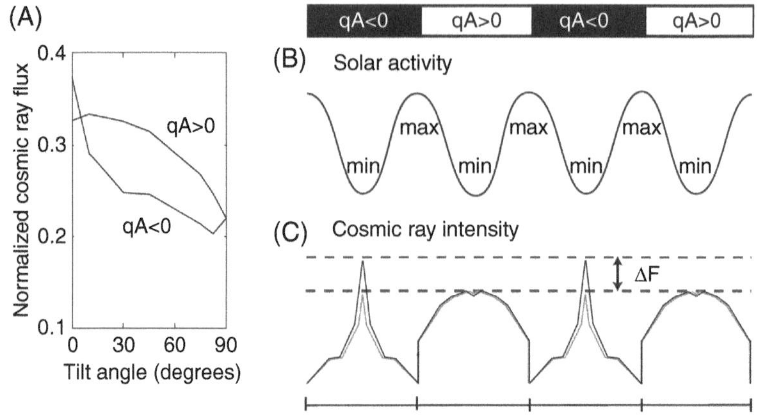

Fig. 5.4 ***Solar modulation of galactic cosmic rays (GCR) largely depends on the tilt angle of heliospheric current sheet (HCS) and the polarity of solar dipole component.*** (A) Tilt angle dependence of Kota and Jokipii (2001), (B) sunspot variation, (C) cosmic ray variation assuming the tilt angle variations within 0–70° (black) or 10–70° (gray) (Kataoka et al., 2012a).

where $f(r,p,t)$ is the phase-space distribution function of charged particles as a function of the momentum p, position r, and time t. $\mathbf{v}$ is the solar wind velocity, and $\tilde{\kappa}$ is the diffusion tensor. The first, second, and third terms express diffusion, advection, and adiabatic cooling. The Parker equation averages over the pitch angles of charged particles, which is applicable only for isotropic distribution. The off-diagonal components of the diffusion tensor stand for the magnetic field curvature and gradient B drift of charged particles. One of the main challenges in modeling GCR modulation is the reliable description of the diffusion tensor throughout the heliosphere.

The magnetic drift as shown in Eqs. (2.12) and (2.13) depends on the particle charge. During the positive magnetic polarity (northward dipole moment) of the Sun, protons drift from heliospheric polar region, while electrons come from the equator, and vice versa during the negative polarity. The tilt angle of HCS also affects the modulation (Kota and Jokipii, 2001). Considering the tilt-angle dependence of magnetic drift protons, the expected flat-top and peaky-top 22-year pattern is graphically summarized in Fig. 5.4 (Miyahara et al., 2009).

The magnetic drift effects can be simulated by solving (Eq. 5.2). The diffusion tensor $\tilde{\kappa}$ is given as a function of r and the magnetic field. Assuming also the time variation of tilted HCS structures through the heliosphere, time-backward SDE simulation reproduces the flat-top and peaky-top 22-year

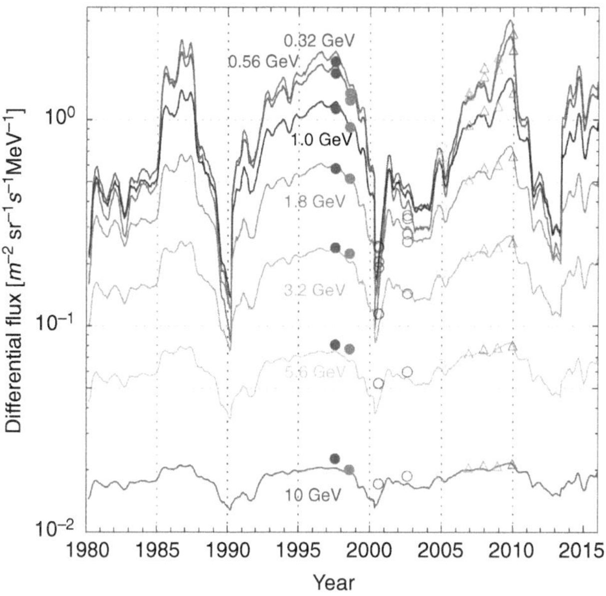

Fig. 5.5 ***Simulated curves and observed data points of cosmic ray flux.*** Important parameters to change the GCR spectrum are tilt angle, magnetic field, and solar wind speed (Miyake et al., 2017).

pattern of GCR as shown in Fig. 5.5. Miyake et al. (2017) provided the space climate forecast (~5-year lead time) of GCR intensity as well as radiation dose for aircrews based on this method.

The above explained solar modulation theory cannot explain the low-energy cosmic rays at ~100 MeV range, which is known as *anomalous cosmic rays* (ACR). The ACR is likely associated with the particle acceleration at the termination shock (Jokipii and Kota, 2000), located in the outer heliosphere (Fig. 1.8).

The information described above gives a basic knowledge on the ambient radiation dose when to book your travel-plan toward Mars in future. After all, it is difficult to fully protect the radiation dose from GCR because of their high energy. The peak-to-peak amplitude of the solar modulation is typically only 20%. During the solar minima of the Maunder Minimum, however, it can have a 40% increase in addition (Kataoka et al., 2012a), which is a possible extreme case when the heliospheric barrier is almost disappeared by the peculiar condition of the flat HCS. High-resolution MHD simulations to resolve the HCS is therefore developed to understand the extreme situations (Matsumoto et al., 2019).

The negative spikes appeared in Fig. 5.1, in 1991 and 2003 for example, are associated with *Forbush decrease* events, as caused by the additional magnetic barrier around the Earth by the passage of CMEs and associated shocks. Although the GCR dose is smaller than average during Forbush decrease events, it is not a good news for interplanetary travelers because SEP events are associated with such CMEs and shocks. Therefore, as another necessary information, there is a worry of episodic radiation dose by SEP events, as shown in the next Section in detail. The radiation dose by SEP can be reasonably protected because the energy is less than that of GCR. However, if we failed to shield, the radiation dose can be critical, exceeding the safety level.

5.2 Predicting solar energetic particles (SEP)

SEP particles are predominantly composed of protons with a minor contribution from helium ions (~10%) and an even smaller part of electrons and heavy ions (~1%). Many other fundamental knowledges of SEP particles have been systematically summarized by Vainio et al. (2009), Desai and Giacalone (2016), and Reames (2017). Following discussions are dedicated to the space weather forecast for a safe travel to and from Mars.

Real-time measured CME speed is practically useful to roughly predict the SEP intensity because there is a positive correlation between the CME speed and >10 MeV SEP intensity, although the error is more than an order of magnitude (Gopalswamy et al., 2004; Takahashi et al., 2016). A better empirical prediction formula was derived combining STEREO observation data (Richardson et al., 2014) as follows:

$$I = 0.013 \exp\left(0.0036V - \frac{\varphi}{2\sigma^2}\right), \tag{5.3}$$

where I is the peak intensity of 14 to 24 MeV proton intensity in units of (MeV cm^{-2} sr s)$^{-1}$, V is the CME speed, and φ is the longitudinal distance assuming the Parker spiral. The Gaussian width σ is 43°. This empirical model reduces the prediction errors, and more than 80% of the predictions are within an order of magnitude of the observed SEP intensity (Richardson et al., 2018).

Emslie et al. (2012) showed that SEP has typically a few percent of the CME kinetic energy measured in the rest frame of the solar wind. Such an energetics is also useful to roughly limit the maximum intensity of SEP events. Similarly, *streaming limit* is an interesting idea, especially for considering the extreme space weather of SEP events. This is a saturation process to generate SEP with propagating interplanetary shocks. Dotted lines in Fig. 5.6 mark the empirical values of the streaming limit.

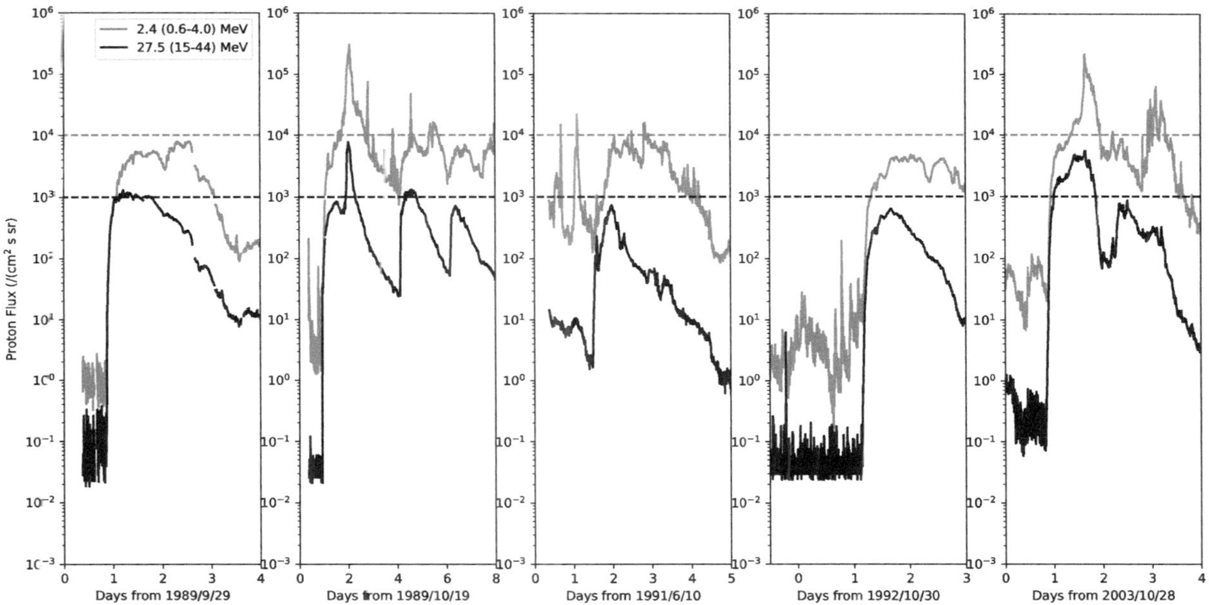

Fig. 5.6 ***Streaming limit of solar energetic particle (SEP) intensity.*** The pioneering work was given by Reames and Ng (1998).

Theoretically, Alfvén waves excited by the energetic particles self-limit the streaming particles away from the shock (Vainio, 2003). The SEP flux of *energetic storm particles* (near-shock observation of shock-accelerated protons) can be beyond the streaming limit (Fig. 5.6). Notifying the peaks of energetic storm particles, associated with shock arrival, is essential to categorize the time profile of SEP observed at different longitudes (Fig. 5.7). The basic idea how to possibly model the varieties of energy spectra of energetic storm particles has been discussed in detail by Tylka et al. (2005).

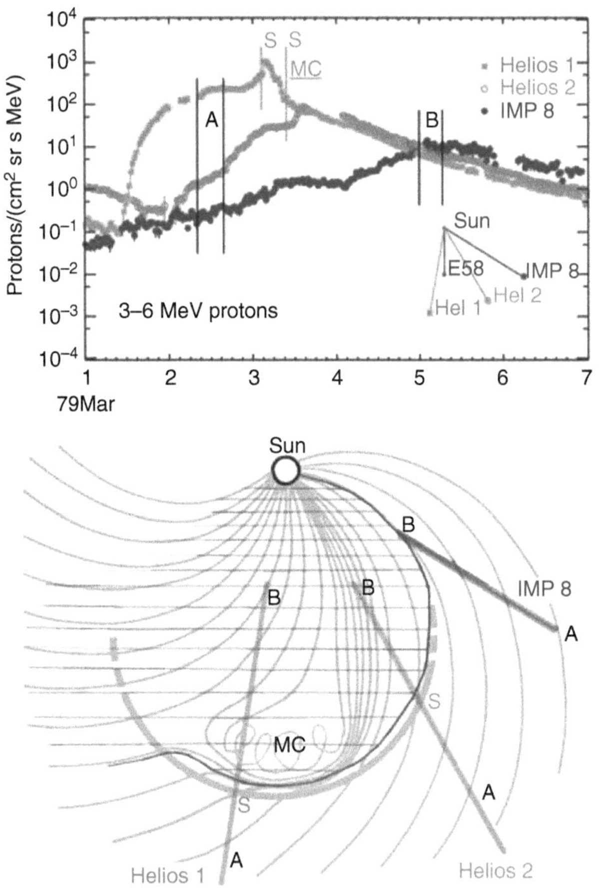

Fig. 5.7 ***Example event showing energetic storm particles (intensity peak associated with shock arrival denoted by "S"), and reservoir (the same energy spectra by three spacecraft, depicted by the shaded region in the bottom panel) (Reames, 2010).***

SEP events tend to have large regions of space, occurring late in the event, where there is little or no spatial gradient in the SEP intensities in longitude, latitude, or radius, which is called *reservoir* or *invariant spectral region* (Fig. 5.7). Reames et al. (2013) presented several examples of the multi-spacecraft observation data from 1977 to 1979 to discuss the importance of the reservoir as the seed spectrum for the further particle acceleration by the second shock passage.

The most energetic population of SEP events are represented by GLE events, which can be modeled and forecasted by the WASAVIES (Kataoka et al., 2014, 2018; Sato et al., 2018b). However, WASAVIES only works for the Earth, because the real-time monitoring data set, such as neutron monitors and GOES, is necessary as the constrain data. For the Moon, we can use the WASAVIES after omitting the magnetosphere module because the Moon is relatively close to the Earth and the 60 R_E distance can be ignored for GeV protons. For Mars and the travel toward Mars, however, the 1D model of WASAVIES does not work anymore, and both 3D SEP propagation model and new constrain data at its position are needed (Fig. 5.8). Therefore, 3D versions such as SPARX are developed (Marsh et al., 2015). As a different approach, an example of the initial experiment of EMMREM framework (Schwadron et al., 2010) was presented for 2003

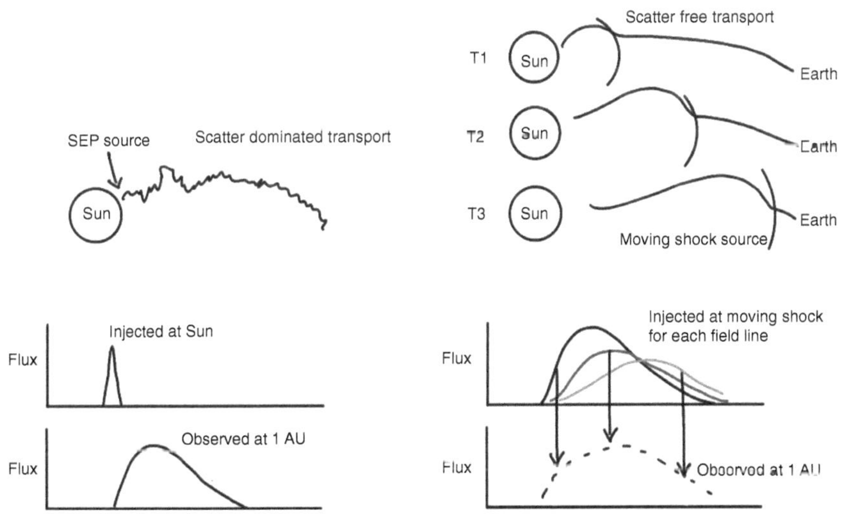

Fig. 5.8 ***Schematic illustrations of the 1D model of WASAVIES for reproducing ground-level enhancement (GLE) events and the 3D model of solar energetic particle (SEP) events, such as SEPMOD (Luhmann et al., 2007).***

Halloween event, where the Earth, Mars, and Ulysses were located along the same Parker spiral at different radial distances (Kozarev et al., 2010).

The relevant proton energy to be predicted is >10 MeV, because assuming the 10-mm thick wall of the spacecraft, we have to shield against an order of 10 MeV protons (Section 5.3). However, the behavior of 10 MeV protons is more complex than that of GeV protons mainly because they are continuously produced in the evolving interplanetary shocks. Note that if we put the solar wind parameters to the Eqs. 2.1 and 2.2, the maximum energy of diffusive shock acceleration becomes an order of 10 MeV. As a recent example of the possible radiation dose, the actual measurements indicate that the 30-day dose limit to the skin, 1.5 Gy (NASA, 2014), would have been approached during the September 2017 SEP event on the lunar surface for a lightly shielded environment (Schwadron et al., 2018).

The probability of suffering from large-amplitude SEP events, including a random occurrence of extreme events, is not negligible for interplanetary travelers. Lockwood and Hapgood (2007) estimated that the probabilities of encountering the sickness-level and fatal-level cumulative dose are ~0.3 and ~0.1 after 600 days beyond the Earth, respectively. It is also not easy to judge when to hide from SEP (Reames, 1999). After all, the solar minimum is therefore better for my journey to Mars, if any chance, because the maximum radiation dose by GCR is known in advance (Kataoka et al., 2012a), and the probability of suffering the large-amplitude SEP is minimized.

The modeling works of the global MHD shock evolution and nonthermal components such as SEP have been more or less separated because of the different scales. As an important step to merge the MHD and SEP, Lario et al. (1998) examined that the injection rate of shock-accelerated protons ranging from 50 keV to 100 MeV is proportional to the downstream–upstream velocity ratio. SOLPENCO is the first model using the injection rate to predict the SEP events associated with traveling shocks (Aran et al., 2005).

Combining the ENLIL model and SEPMOD model, Luhmann et al. (2018) illustrated how the event similarities at three observer sites (STEREO-A, Earth, Mars) can be understood from their relative positions and connectivity to interplanetary shocks. Luhmann et al. (2017) examined several other examples using STEREO spacecraft to see the basic performance. The forward modeling frameworks such as SEPMOD (Fig. 5.8), are useful as physics-based prediction for the conditions on Mars. Similar approach combining the time-dependent 3D MHD model EUHFORIA and the 3D propagation of SEP inside is ongoing (Wijsen et al., 2019). It is promising that major SEP events will be predictable by physics-based models, since we are improving our understanding of realistic 3D solar wind evolutions, including CMEs and interplanetary shocks. The shock-acceleration of solar protons can be calculated from the solar corona to the inner heliosphere, and the propagation and diffusion of the accelerated protons can also be calculated. In future, the operational solar wind models will therefore be continuously improved to have nonthermal SEP components during disturbed conditions and the reliability will be validated by independent users.

5.3 The Moon

The Moon has been a miracle player for centuries to help understanding the space climate because the Moon occasionally masks the solar disk, and the total solar eclipse has been giving us fruitful hints on solar corona structures as clearly depicted in Fig. 5.9. However, it was after the space age that we correctly understood the solar cycle evolution of the beautiful pattern of coronal emissions and their relevance to our modern society, as introduced in the previous chapters.

The Moon is a neighborhood astronomical body we can visit, footing at the surface. It takes only a second by the light speed, and only a week for

Fig. 5.9 ***A drawing of the 1878 solar eclipse by E. L. Trouvelot.*** The Trouvelot Astronomical Drawings. New York, C. Scribner's Sons, 1881–1882.

Table 5.1 The largest satellites in the solar system.

Satellite	Radius	Orbital radius
The Moon	1737 km	60 R_E
Io	1818 km	6.0 R_J
Europa	1561 km	9.6 R_J
Ganymede	2634 km	15.3 R_J
Callisto	2408 km	26.9 R_J
Titan	2575 km	20 R_S

The Moon, four Gallileo satellites of Jupiter, and a giant satellite of Saturn are shown. R_E is 6378 km, R_J is 71,492 km, and R_S is 60,330 km.

a round trip of humans to the Moon. The Moon is surprisingly big with radius of ~1/4 of the Earth and the mass of 1% of the Earth. The Moon is the fifth largest satellite in the solar system (Table 5.1). All of these satellites are in extreme space weather environment. For example, the largest satellite is Ganymede, which has the global magnetic field, forming Alfvén wing in the magnetosphere of Jupiter (Kivelson et al., 1997).

The orbital radius of the Moon is 60 R_E, spending about 20% of its orbit in the magnetosphere of the Earth (Harnett et al., 2013; Terada et al., 2017). The Moon surface is a dangerous environment, due to the absence of a global magnetic shield and atmospheric shelter against cosmic rays. Building new Moon stations and moon-orbiting stations for various purposes are planned as the start-point to the journey to Mars and beyond. The buildings can be completely covered using lunar regolith to make the thick shelter, while underground life is another option to avoid unnecessary radiation (Naito et al., 2020).

Crustal magnetic anomalies have been found on the Moon (Fig. 5.10), and their magnetic field strength is strong enough to interact with solar wind particles (Futaana et al., 2013; Saito et al., 2012; Bhardwaj et al., 2015; Bamford et al., 2016), although it is still weak and not relevant to shield against energetic particles, such as SEP and GCR. The interaction of the solar wind with the Moon was illustrated in Fig. 5.11. At 30 km altitude, the magnetic field intensity is less than 10 nT, while it is typically less than 100 nT at the surface. Among several magnetic anomaly regions, Gerasimovic anomaly has the highest intensity of 718 nT (Tsunakawa et al., 2015).

To evaluate the radiation shielding against charged particles, the Bethe–Bloch equation (e.g., Ziegler, 1999) is useful to analytically estimate the energy deposition, that is, energy loss per unit distance. Monte-Carlo simulation code PHITS is an alternative option to estimate the

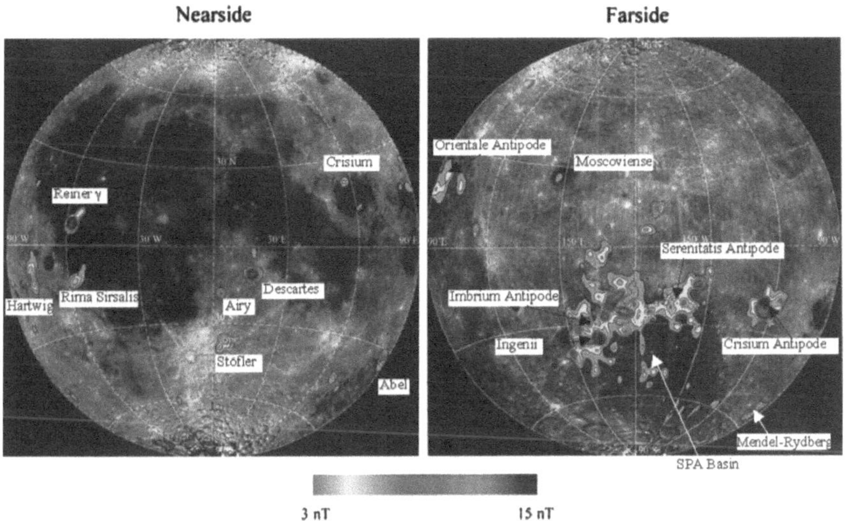

Fig. 5.10 *Locations and intensities of magnetic anomaly regions at 30 km altitude (Hood, 2011).*

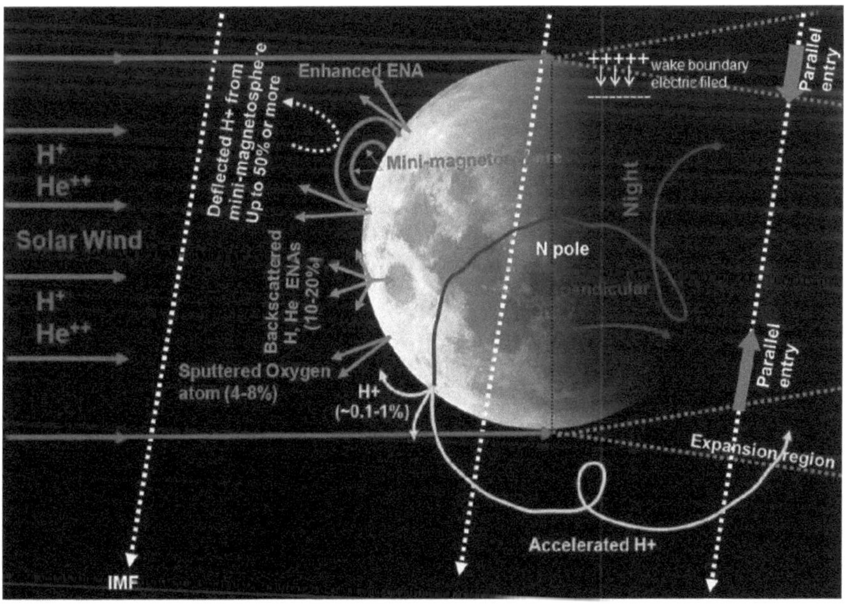

Fig. 5.11 *Schematic illustration of the solar wind interaction with the Moon (Bhardwaj et al., 2015).*

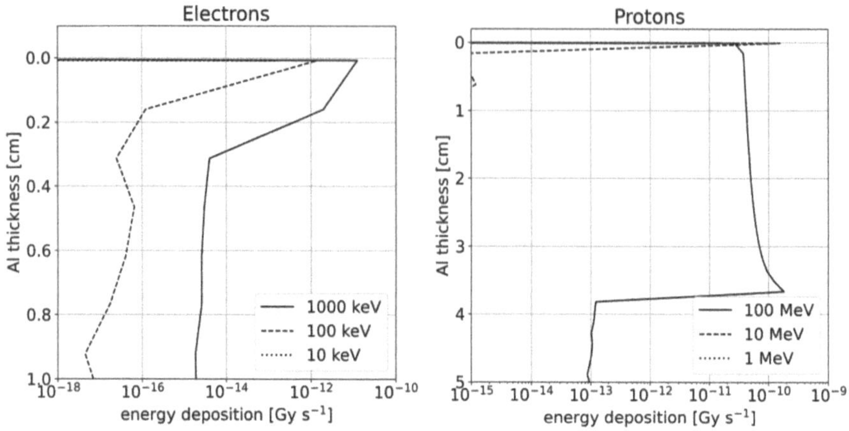

Fig. 5.12 ***Example of PHITS simulation to estimate the energy deposition of (left) electrons and (right) protons penetrating Al sheet.***

energy deposition. For example, imagine how the spacesuit or spacecraft is vulnerable to the penetration of energetic electrons or protons: X-rays are fully blocked by 1 g/cm^2 or 3.7 mm thick Al spacesuits. As shown in Fig. 5.12, Al sheets with 2 and 3 mm thickness stop 0.1 and 1.0 MeV electrons, respectively, while 2 mm and 4 cm thick Al sheets stop 10 and 100 MeV protons, respectively.

In a similar manner, the radiation posed on the Moon surface is recorded on the regolith, and high-energy particles affect deep regolith. The depth profile of the cosmogenic nuclei, such as that of ^{26}Al, on the Moon's surface therefore provides an important hint to estimate the upper limit of the SEP contribution in the past million years, although there is no time resolution. The discontinuity of depth curve provides a hint to distinguish the SEP (soft) and GCR (hard). For example, the occurrence rate of 3×10^{34} erg super-flare SEP event is estimated to be one in million years (Poluianov, 2019). It is also interesting to note that the depth profile of temperature in the lunar regolith may also be useful to understand the past solar activity (Miyahara et al., 2008).

Note also that large SEP events charge the polar lunar regolith, particularly in the permanently shadowed region, where the electrical conductivity is low due to the low temperature (Fig. 5.13). The resultant dielectric breakdown can then change the porosity and albedo of the lunar regolith (Jordan et al., 2015, 2017).

The radiation risk must be quantitatively evaluated for Moon workers. The absorbed dose D is the energy absorbed by unit mass due to ionizing

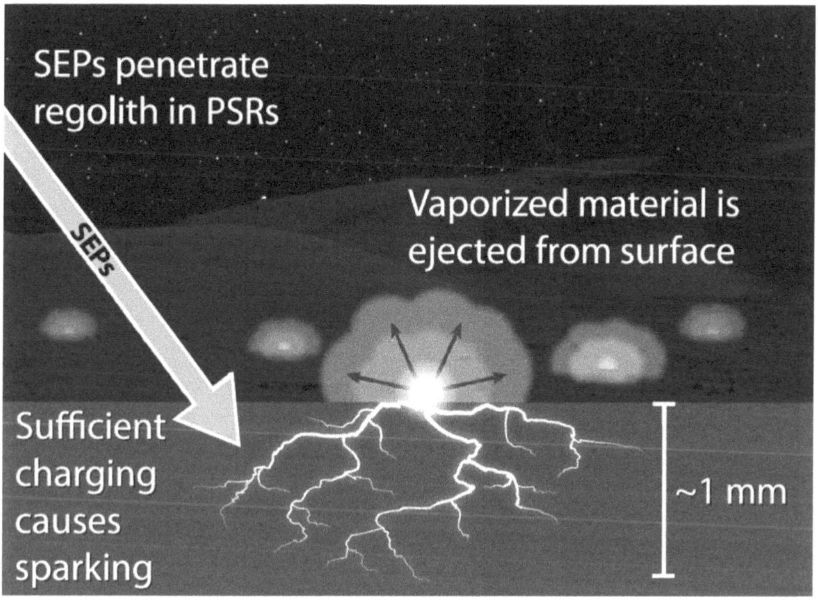

Fig. 5.13 ***Possible impact of solar energetic particles (SEP) on lunar regolith (Jordan et al., 2017).***

radiation. The SI unit of D is Gray, defined as J/kg. The dose equivalent H indicates the risk of occurrence of biological effects due to the absorbed dose D, which is defined by $H = Q \times D$, where the relative biological effectiveness factor Q accounts for the radiation type and characteristics of the affected body organ. The SI unit of H is Sievert.

The probability of cancer death increases at >100 mSv/year. A CAT scan at a hospital gives the exposure of 10 to 40 mSv in one test, while a single chest X-ray gives only 0.1 mSv. It is also worthwhile to note here that the passage of radiation belts of the Earth causes the radiation dose of a few mSv (Märki, 2020). The possible risks and radiation effects are summarized in Table 5.2.

The Apollo astronauts were lucky. Apollo 16 astronauts returned to the Earth on April 27, 1972. On August 4, 1972, one of the largest SEP events occurred. If this SEP event occurred months earlier, it could have exposed the Apollo 16 astronauts on the lunar surface to far higher radiation levels than allowable: Astronauts protected only by a space suit would have received more than 10 Gy radiation to the skin and 2 Gy radiation to the bone marrow (Townsend, 2020).

Table 5.2 Radiation effects in humans after whole body irradiation (Hellweg and Baumstark-Khan, 2007).

Dose (Sv)	Effect	Outcome
~0.2	First evidence of increased cancer risk as late effect	
<0.25	No obvious direct clinical effects	
>0.5	Nausea, vomiting	No early death anticipated.
3–5	Bone marrow syndrome	Death rate for this syndrome peaks at 30 days, but continues out to 60 days. Death occurs from sepsis.
5–12	Gastrointestinal tract syndrome	Deaths from this syndrome occur between 3 and 10 days post exposure. Death occurs from sepsis.
>20	Central nervous system syndrome	No survivors expected.

GCR causes radiation exposure of several hundred mSv per year on average on the Moon and Mars. The GCR dose rate depends on the solar cycle phase as discussed in Section 5.1. On the Moon, it is estimated to be ~250 and ~900 μSv/day in the solar maximum and minimum, respectively, where irons have a larger contribution than protons (Sato et al., 2018a). On Mars, it is expected to be ~180 and ~500 μSv/day in the solar maximum and minimum, respectively, where neutrons have larger contribution than protons (Sato et al., 2018a).

The solar cycle variation of GCR was observed around the Moon, by the Cosmic Ray Telescope for the Effects of Radiation (CRaTER) onboard the Lunar Reconnaissance Orbiter. Zeitlin et al. (2019) showed that the GCR proton flux in 2018 was approaching that in the previous deep minimum in 2009. Recently, Zhang et al. (2020) reported that their Lunar Lander Neutrons and Dosimetry experiment onboard Chang'E 4 lander measured an average dose equivalent of 1369 μSv/day on the lunar surface. For the same time period, the dose equivalent at the ISS was 731 μSv/day with GCR contributions of 523 μSv/day. The additional ~208 μSv/day is due to protons in the South Atlantic Anomaly. Therefore, the daily GCR dose equivalent on the lunar surface is ~2.6 times higher than that inside the ISS (Zhang et al., 2020). Such a direct observation at the Moon surface is important. We know on the Earth, watching a fine-resolution weather forecast of heavy rain map is useful to decide to move or not, particularly in urgent situations. In future, space weather stations will be constructed at the Moon to provide a detailed radiation hazard map working real-time.

Other than the radiation dose, electrostatic charging is the space weather of great concern for objects and astronauts on the lunar surface. For example, the electrical grounding in shadowed craters of the Moon is challenging, considering the triboelectric charge accumulation when using a drill under the environmental currents from ambient solar wind plasma (Rhodes et al., 2020). Moreover, spacesuit surface potential is dependent on both ambient plasma condition and dust coverage, and the electrostatic discharge may pose a serious risk to astronauts (Chou et al., 2019). The lunar surface charging has been examined by the Lunar Prospector experiments during SEP events (Halekas et al., 2007; Halekas et al., 2009) where indications of negative potentials of up to 4.5 kV were identified. It is suggested that SEP electrons may play an important role to drive the high-voltage events (Borovsky and Delzanno, 2021).

5.4 Mars

After taking a giant leap by setting foot on the Moon for the first time in July 1969, the next giant leap to Mars is about to be taken. In February 2018, SpaceX launched a car to the Mars orbit. In March 2021, a new rover Perseverance landed on Mars, and in April, the first helicopter flew over Mars. The main purpose of these missions is to search for life on Mars.

As summarized in Table 5.3, Mars has a 1.4 to 1.6 AU elliptical orbit, which is a unique characteristic compared to the Earth and Venus. The distance between Mars and the Earth varies from 56 to 400 million km. The time required for radio communication also changes from 3 min to

Table 5.3 Characteristics of terrestrial planets.

Parameter	Venus	Earth	Mars
Radius	6052 km	6378 km	3396 km
Orbital radius	0.72 AU	1 AU	1.4–1.6 AU
Rotation period	243 days	24 hours	24.6 hours
Surface gravity	8.9 m/s^2	9.8 m/s^2	3.7 m/s^2
Escape velocity	10 km/s	11 km/s	5 km/s
Surface temperature	740 K	288 K	210 K
Surface pressure	92,000 hPa	1000 hPa	7 hPa
Composition	96% CO_2 3.5% N_2	78% N_2 21% O_2	95% CO_2 2.7% N_2

22 min, respectively. The radio communication is disrupted during the solar conjunction when Mars moves behind the Sun from the perspective of the Earth. For example, in August 2002, during the conjunction, no data from Martian Radiation Environment Experiment (MARIE) instrument onboard Mars Odyssey were received for approximately two weeks (Atwell et al., 2003). MARIE aimed to assess the radiation environment at Mars to determine the radiation risk that the astronauts on a Mars mission may encounter, and it stopped working properly on October 28, 2003 during the Halloween event (Crosby, 2007). During the latest solar conjunction in 2021, NASA missions such as Perseverance rover, Ingenuity helicopter, Curiosity rover, InSight lander, Odyssey, Mars Reconnaissance Orbiter, and the Mars Atmosphere and Volatile Evolution Mission (MAVEN), stopped sending commands and the data between October 2 and 16, 2021.

The Mars is the planet of dust storm (Sagan and Pollack, 1969). It is suggested that the dust causing zodiacal light might even originate in Mars, or its two satellites, the Phobos and the Deimos, based on the direct measurement by the solar array of Juno spacecraft as a dust impact target (Jorgensen et al., 2021). During the dust storm, however, we do not have to worry about the windy environment to knock down the rockets or buildings, for example, because the air pressure is less than 1% of the Earth (Table 5.3). There are many other obstacles that must be overcome to realize our visit on Mars. Nevertheless, Mars is one of the most interesting planets to walk on because it was possibly a habitable planet. We will understand why it turned to a dry planet by closely looking at the surface of Mars, to predict our future.

The related popular topic is the Faint-Young-Sun paradox, which was introduced by Sagan and Mullen (1972). Based on standard stellar evolution models, the young Sun's luminosity was about 30% lower than its current luminosity. The surface equilibrium temperature of the Earth and Mars at that time must then be well below the freezing point of water, but geological evidence supports the existence of liquid water both at Earth and Mars. Therefore, we need to introduce some special heating process, such as runaway greenhouse effects, to increase the average surface temperature of the Earth and Mars.

The geological evidence of ancient fluvial activity on Mars surface suggests the warm and wet climate under a thick atmosphere (Pollack et al., 1987). Against the warn and wet scenario, however, there is cold and icy scenario by carefully discussing the greenhouse warming effects and the resultant mean surface temperature around the freezing point of water (Forget et al. 2013; Wordsworth et al., 2013; Ramirez et al., 2014).

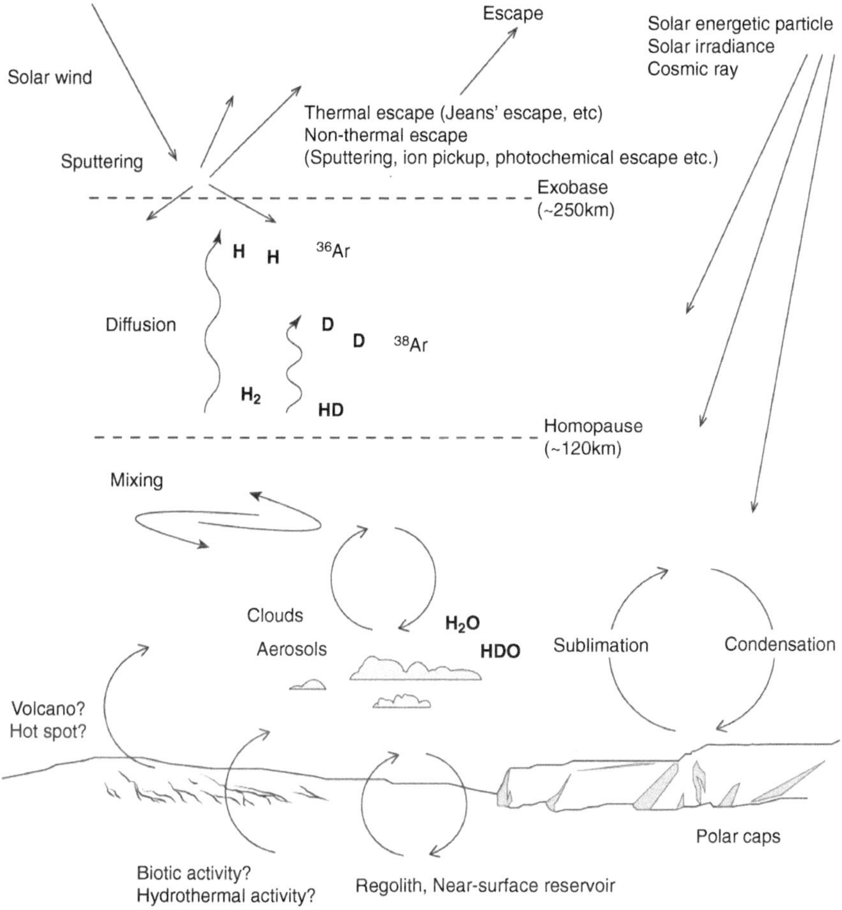

Fig. 5.14 ***Schematic illustration of the Marian atmospheric evolution with various key processes.*** Both energetic photons and energetic particles affect the atmosphere to change the Martian environment (Nakagawa, 2019). *Courtesy of H. Nakagawa.*

At present, the Martian atmosphere is directly exposed to the solar wind and experiences various atmospheric escape processes, which can alter the basic atmospheric evolution (Fig. 5.14). There is a hypothesis that the solar wind or CMEs could blow out the ancient thick atmosphere of Mars (Jakovsky et al., 2017). Although the present Mars does not have a global magnetic field, it might have had a global magnetic field ~4 billion years ago (Lillis et al., 2013), when the solar activity was likely extremely high with more than an order of difference in ultraviolet intensity (Ribas et al., 2005; Tu et al., 2015), dense and high-speed solar wind (Wood et al., 2002, 2005), and the powerful and frequent CMEs (Terada et al., 2009). Actively

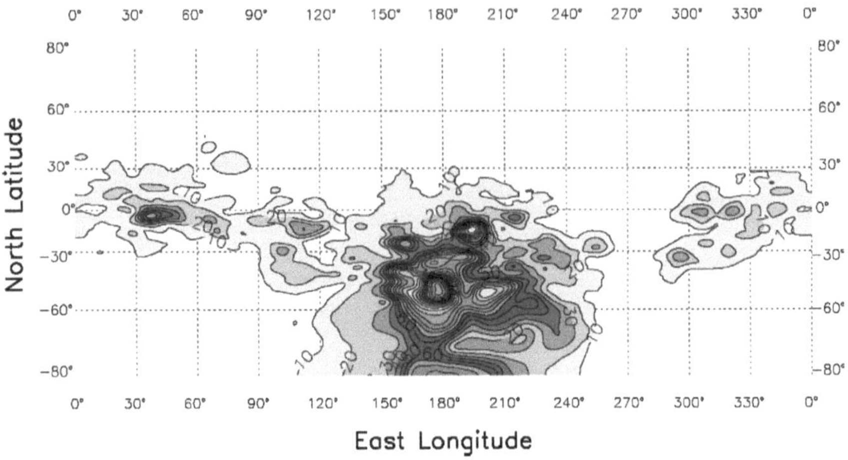

Fig. 5.15 ***The magnetic field strength in nano Tesla at ~400 km altitude as observed by Mars Global Surveyor (Hood et al., 2005).***

discussed topics include the essential role of the relatively weak global magnetic field (Weiss et al., 2008) to protect the atmosphere of Mars from ion escape (Sakata et al., 2020; Sakai et al., 2021).

The present Mars has a remnant magnetism that was locked in the crust, particularly in the southern hemisphere about 4 billion years ago (Fig. 5.15), while the rest of the planet has a negligible intrinsic magnetic field. Even though the induced magnetosphere (Luhmann et al., 2004) exists, the GMD or radiation belts are not important for future missions due to the absence of magnetospheric current generator and trapped energetic electrons and protons, unless artificial magnetospheres are created there (Bamford et al., 2021). The possible roles of the induced magnetosphere for the atmospheric loss occurring at present Mars have also been actively discussed (Egan et al., 2019; Gronoff et al., 2020).

In terms of space radiation, living on the surface of Mars is a situation similar to the flying-high condition on the Earth. In other words, the atmospheric pressure at the surface of Mars is comparable to that at the Earth stratosphere. As a radiation shelter, the Earth atmosphere is equivalent to 1 m deep water, but the atmosphere of Mars corresponds to only 1 cm deep water.

Fig. 5.16 provides an example of PHITS simulation using the Mars Global Reference Atmospheric Model 2001 (Mars-GRAM 2001) (Justus and Johnson 2001; Justus et al. 2006). It is clear that more than >150 MeV protons can penetrate to the Martian surface (Hassler et al., 2014), while low energy SEP components are absorbed in the atmosphere.

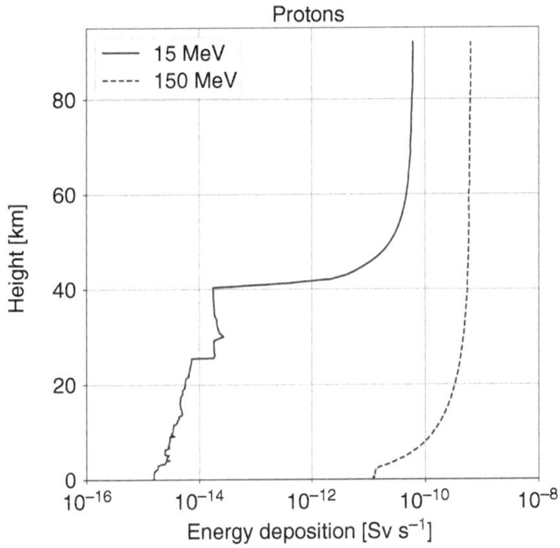

Fig. 5.16 ***Energy deposition rates of 15 MeV and 150 MeV protons in the Mars atmosphere, simulated by PHITS and Mars-GRAM 2001 model.***

The ready-to-go functions to calculate the radiation dose for a given energy spectra of SEP or GCR on the Martian surface has also been provided (Guo et al., 2019). Hassler et al. (2014) reported that the 500-day surface stay on Mars yields a 0.3 Sv dose equivalent, which is comparable to the 180-day transit to Mars. Note also that, for practical applications, we cannot neglect the geographical features because the Mars has the largest volcanoes (>20 km altitude) in the solar system, called Olympus Mons and a large crater (−4 km altitude) called Hellas Planitia.

On the Earth, ozone depletion is the clear outcome from the atmospheric ionization by SEP (Jackman et al., 2005), as introduced in Section 2.3. On Mars, similar effect on the atmospheric chemistry has been actively discussed. For example, Airapetian et al. (2016) suggested that HCN, N_2O and H_2CO may be produced by ancient strong SEP events, which can be constituents for ancient life and greenhouse effects on early Mars.

During the September 2017 SEP event, the largest increase in dose rate (Fig. 5.17) and energetic proton flux (Fig 5.18) were observed by the Radiation Assessment Detector (RAD) on the Mars Science Laboratory Curiosity rover (Ehresmann et al., 2018; Hassler et al., 2018). Hassler et al. (2014) reported the RAD observation during the second largest SEP event on April 11, 2013, just after the solar maximum of solar cycle 24.

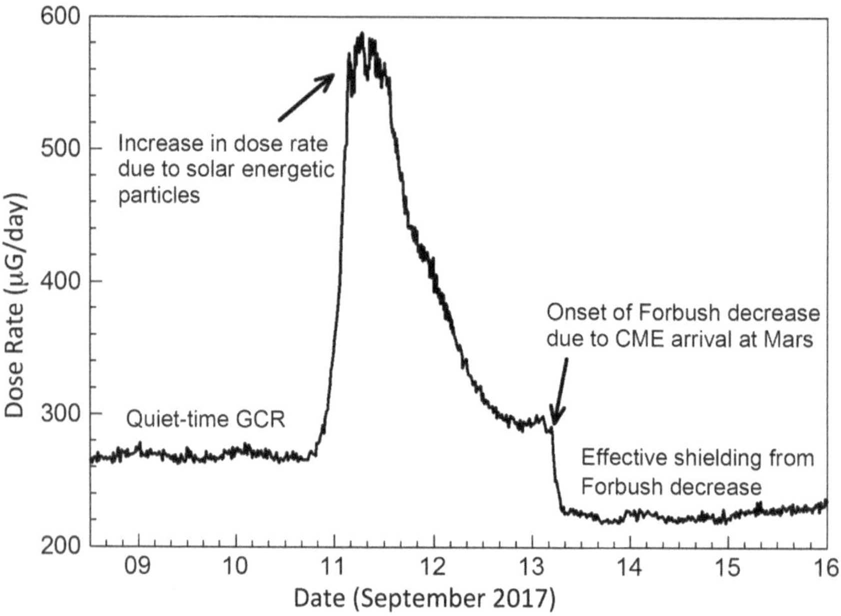

Fig. 5.17 ***Radiation dose rate observation during the September 2017 SEP event.*** The dose rate increased above the background GCR levels by a factor of 2. A Forbush decrease was also observed with a sudden drop of about 15% in the dose rate (Hassler et al., 2018).

Mars diffuse aurora (Fig. 5.19) was also caused by the SEP event in September, 2017 (Schneider et al., 2018). The altitude profile of CO_2 emission is extended down to ~60 km altitude. Such a deep atmospheric ionization requires energetic electrons up to 100 keV (Gerard et al., 2017; Haider and Masoom, 2019) and/or MeV protons (Nakamura et al., 2022). Other than the September 2017 event (Lee et al., 2018), Lee et al. (2017) summarized recent examples of the detailed time variation of the solar wind parameters and energy spectra for several pre-2017 SEP events, during the declining phase of solar cycle 24.

During the lengthy travel to Mars, at least more than 6-month trip, spaceship wall thickness thus matters as the atmospheric shelter substitute for humans. Special care must also be paid to other possible damages on the spaceship control. On July 23, 2012, the large SEP event degraded >10% solar panel output of Akatsuki spacecraft heading to Venus. Such an episodic and permanent damage in the essential electronics may also occur during the journey to Mars in future.

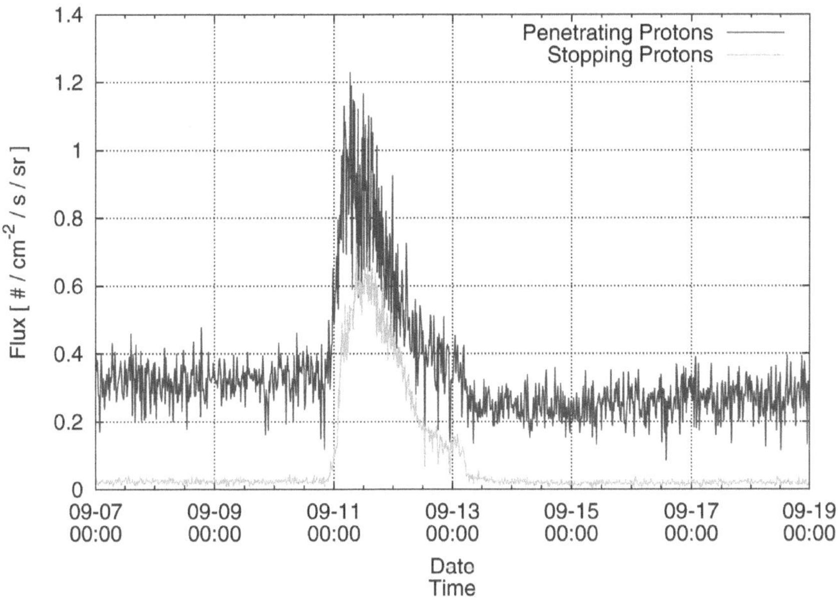

Fig. 5.18 ***Integrated proton fluxes versus time at the surface of Mars.*** The penetrating proton flux of >135 MeV shows increase by a factor of 3–4, while the stopping proton flux <100 MeV shows increase by a factor of 30, compared to the quiet time (Ehresmann et al., 2018).

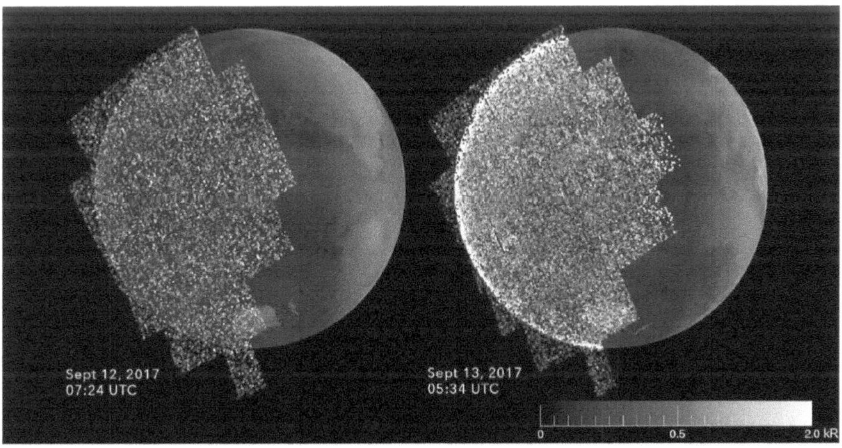

Fig. 5.19 ***Mars global diffuse aurora caused by the SEP event in September, 2017, as observed by MAVEN/IUVS (Schneider et al., 2018).***

References

Adriani, O. et al. (CALET Collaboration), 2019. Direct measurement of the cosmic-ray proton spectrum from 50 GeV to 10 TeV with the calorimetric electron telescope on the International Space Station. Phys. Rev. Lett. 122, 181102.

Akasofu, S.-I., Chapman, S., 1962. Large-scale auroral motions and polar magnetic disturbances III: the aurora and magnetic storm of 11 February, 1958. J. Atmos. Terr. Phys. 24, 596–785.

Akasofu, S.-I., 1964. The development of the auroral substorm. Planet. Space Sci. 12 (4), 273–282. https://doi.org/10.1016/0032-0633(64)90151-5.

Akasofu, S.-I., Kamide, Y., 2005. Comment on "The extreme magnetic storm of 1-2 September 1859" by B.T. Tsurutani, W.D. Gonzalez, G.S. Lakhina, and S. Alex. J. Geophys. Res. 110, A09226. doi:10.1029/2005JA011005.

Airapetian, V., Glocer, A., Gronoff, G., et al., 2016. Prebiotic chemistry and atmospheric warming of early Earth by an active young Sun. Nature Geosci. 9, 452–455. https://doi.org/10.1038/ngeo2719.

Alken, P., Thebault, E., Beggan, C.D., Amit, H., Aubert, J., et al., 2021. International geomagnetic reference field: the thirteenth generation. Earth Planets Space 73, 49. https://doi.org/10.1186/s40623-020-01288-x.

Allen, J., Frank, L., Sauer, H., Reiff, P., 1989. Effects of the March 1989 solar activity. EOS 70 (46), 1479, 1486–1488.

Alfvén, H., 1942. Existence of electromagnetic-hydrodynamic waves. Nature 150, 405–406.

Amano, T., Seki, K., Miyoshi, Y., Umeda, T., Matsumoto, Y., Ebihara, Y., Saito, S., 2011. Self-consistent kinetic numerical simulation model for ring current particles in the Earth's inner magnetosphere. J. Geophys. Res. 116, A02216. doi:10.1029/2010JA015682.

Anderson, P.C., McKenzie, D.L., Datlowe, D.W., Hawley, J.D., Petrinec, S.M., Schulz, M., Larson, D.E., 2000. Polar cap X-rays and electrons under low density solar wind conditions: coordinated PIXIE and DMSP observations on 11 May 1999. Geophys. Res. Lett. 27 (24), 4021–4024. https://doi.org/10.1029/2000GL011983.

Akasofu, S.-I., 1981. Energy coupling between the solar wind and the magnetosphere. Space Sci. Rev. 28, 121–190.

Anderson III, C.W., Lanzerotti, L.J., MacLennan, C.G., 1974. Outage of the L4 system and the geomagnetic disturbances of 4 August 1972. Bell Syst. Tech. J. 53, 1817–1837. https://doi.org/10.1002/j.1538-7305.1974.tb02817.x.

Araki, T., 1994. A physical model of the geomagnetic sudden commencement. In: Engebretson, M.J., Takahashi, K., Scholer, M. (Eds.). Solar Wind Sources of Magnetospheric Ultra-Low-Frequency Waves, Geophysical Monograph, 81, pp. 183–200.

Araki, T., Fujitani, S., Emoto, M., Yumoto, K., Shiokawa, K., Ichinose. H. Luehr, T., Orr, D., Milling, D., Singer, H., Rostoker, G., Stunomura, S., Yamada, Y., Liu, C.F., 1997. Anomalous sudden commencement on March 24, 1991. J. Geophys. Res. 102 (A07), 4075–4086.

Araki, T., 2014. Historically largest geomagnetic sudden commencement (SC) since 1868. Earth Planets Space 66, 164. doi:10.1086/s40623-014-0164-0.

Aran, A., Sanahuja, B., Lario, D., 2005. Fluxes and fluences of SEP events derived from SOLPENCO. Ann. Geophys. 23, 3047–3053.

Arge, C.N., Pizzo, V.J., 2000. Improvement in the prediction of solar wind conditions using near-real time solar magnetic field updates. J. Geophys. Res. 105, 10465–10480.

Arge, C.N., Luhmann, J.G., Odstrcil, D., Schrijver, C.J., Li, Y., 2004. Stream structure and coronal sources of the solar wind during the May 12th, 1997 CME. J. Atmos. Sol.-Terr Phys. 66, 1295–1309.

Asvestari, E., Pomoell, J., Kilpua, E., Good, S., Chatzistergos, T., Temmer, M., Palmerio, E., Poedts, S., Magdalenic, J., 2021. Modelling a multi-spacecraft coronal mass ejection encounter with EUHFORIA. Astron. Astrophys. 652 (2021), A27. https://doi.org/10.1051/0004-6361/202140315.

Atwell, W., 2003. The Martian Radiation Environment Experiment (MARIE) on the 2001 Mars Odyssey Spacecraft. AIP Conf. Proc. 654, 1026–1033. https://doi.org/10.1063/1.1541398.

Bagenal, F., 1992. Giant planet magnetospheres. Annu. Rev. Earth Planet. Sci. 20, 289–328.

Baker, D.N., et al., 2008. Severe Space Weather Events-Understanding Societal and Economic Impacts. Natl. Acad. Press., Washington, D.C.

Baker, D.N., Li, X., Pulkkinen, A., Ngwira, C.M., Mays, M.L., Galvin, A.B., Simunac, K.D.C., 2013a. A major solar eruptive event in July 2012: Defining extreme space weather scenarios. Space Weather 11, 585–591. doi:10.1002/swe.20097.

Baker, D.N., et al., 2013b. A long-lived relativistic electron storage ring embedded in Earth's outer Van Allen Belt. Science 340 (6129), 186–190. doi:10.1126/science.1233518.

Baker, D.N., Lanzerotti, L.J., 2016. Resource letter SW1: space weather. Am. J. Phys. 84 (2016), 166. https://doi.org/10.1119/1.4938403.

Baker, D.N., Punasyuk, M.I., 2017. Discovering Earth's radiation belts. Phys. Today December, 46-51.

Baker, D.N., Erickson, P.J., Fennell, J.F., et al., 2018. Space weather effects in the Earth's radiation belts. Space Sci. Rev. 214, 17. https://doi.org/10.1007/s11214-017-0452-7.

Baker, D.N., Bodeau, M., 2021. Effects of space radiation on contemporary space-based systems I: single event upsets, spacecraft charging, degradation of electronics, and attenuation on fiber cabling. In: Coster, A.J., Erickson, P.J., Lanzerotti, L.J. (Eds.). Space Weather Effects and Applications. Geophysical Monograph Series, 262, pp. 3–11.

Bamford, R.A., et al., 2016. 3D PIC simulations of collisionless shocks at lunar magentic anomalies and their role in forming lunar swirls. Astrophys. J. 830 (2), 146.

Bamford, R.A., Kellett, B.J., Green, J.L., Dong, C., Airapetian, V., Bingham, R., 2022. How to create an artificial magnetosphere for Mars. Acta Astronaut. 190, 323–333. https://doi.org/10.1016/j.actaastro.2021.09.023.

Beer, J., Tobias, S., Weiss, N., 1998. An active Sun throughout the Maunder Minimum. Sol. Phys. 181, 237–249. https://doi.org/10.1023/A:1005026001784.

Belcher, J.W., Davis, L., 1971. Large-amplitude Alfven waves in the interplanetary medium, 2. J. Geophys. Res. 76 (16), 3534–3563. doi:10.1029/JA076i016p03534.

Bell, A.R., 1978. The acceleration of cosmic rays in shock fronts – I. Monthly Notices of the Royal Astronomical Society 182 (2), 147–156, https://doi.org/10.1093/mnras/182.2.147.

Berger, T., et al., 2018. The solar particle event on 10 September 2017 as observed onboard the International Space Station (ISS). Space Weather 16, 1173–1189. https://doi.org/10.1029/2018SW001920.

Bieber, J.W., Clem, J., Evenson, P., Pyle, R., Saiz, A., Ruffolo, D., 2013. Giant ground level enhancement of relativistic solar protons on 2005 January 20. I. Spaceship Earth Observations. Astrophys. J. 771 (92), 13. doi:10.1088/0004-637X/771/2/92.

Blake, J.B., Gussenhoven, M.S., Mullen, E.G., Fillius, R.W., 1992. Identification of an unexpected space radiation hazard. IEEE Trans. Nucl. Sci. 39 (6), 1761–1764.

Blandford, R.D., Ostriker, J.P., 1978. Particle acceleration by astrophysical shocks. Astrophys. J. 221, L29–L32.

Blum, L.W., Breneman, A.W., 2020. From Loss in the Magnetosphere to Particle Precipitation in the Atmosphere. Chapter 3 - Observations of radiation belt losses due to cyclotron wave-particle interactions. In: Jaynes, A.N., Usanova, M.E. (Eds.), The Dynamic Loss of Earth's Radiation Belts, Elsevier, pp. 49–98, https://doi.org/10.1016/B978-0-12-813371-2.00003-2.

Bodeau, M., Baker, D.N., 2021. Effects of space radiation on contemporary space-based systems II: spacecraft internal and external charging and discharging effects. In: Coster,

A.J., Erickson, P.J., Lanzerotti, L.J. (Eds.), Space Weather Effects and Applications, Geophysical Monograph Series, 262, pp. 13–61.
Bolduc, L., 2002. GIC observations and studies in the Hydro-Quebec power system. J. Atmos. Sol. Terr. Phys. 64, 1793–1802.
Bonner IV, L.R., Schultz, A., 2017. Rapid prediction of electric fields associated with geomagnetically induced currents in the presence of three-dimensional ground structure: Projection of remote magnetic observatory data through magnetotelluric impedance tensors. Space Weather 15, 204–227. doi:10.1002/2016SW001535.
Borovsky, J. E., Delzanno, G. L., 2021. Do Impulsive Solar-Energetic-Electron (SEE) Events Drive High-Voltage Charging Events on the Nightside of the Moon? https://doi.org/10.3389/fspas.2021.655333
Boteler, D.H., 2006. The super storms of August/September 1859 and their effects on the telegraph system. Adv. Space Res. 38, 159–172.
Boteler, D.H., Pirjola, R.J., 2017. Modeling geomagnetically induced currents. Space Weather 15, 258–276. doi:10.1002/2016SW001499.
Boteler, D.H., 2019. A 21st century view of the March 1989 magnetic storm. Space Weather 17. doi:10.1029/2019SW002278.
Bothmer, V., Rust, D., 1997. The field configuration of magnetic clouds and the solar cycle. In: Crooker, N., Joselyn, J.A., Feynman, J. (Eds.), Coronal Mass Ejections. https://doi.org/10.1029/GM099p0139.
Brehm, N., Bayliss, A., Christl, M., et al., 2021. Eleven-year solar cycles over the last millennium revealed by radiocarbon in tree rings. Nat. Geosci. 14, 10–15. https://doi.org/10.1038/s41561-020-00674-0.
Brodrick, D., Tingay, S., Wieringa, M., 2005. X-ray magnitude of the 4 November 2003 solar flare inferred from the ionospheric attenuation of the galactic radio background. J. Geophys. Res. 110, A09S36. doi:10.1029/2004JA010960.
Bruno, A., Christian, E.R., de Nolfo, G.A., Richardson, I.G., Ryan, J.M., 2019. Spectral analysis of the September 2017 solar energetic particle events. Space Weather 17, 419–437. doi:10.1029/2018SW002085.
Burlaga, L.F., Skoug, R.M., Smith, C.W., Webb, D.F., Zurbuchen, T.H., Reinard, A., 2001. Fast ejecta during the ascending phase of solar cycle 23: ACE observations, 1998?1999. J. Geophys. Res. 106 (A10), 20957–20977. doi:10.1029/2000JA000214.
Burlaga, L.F., Ness, N.F., Stone, E.C., 2013. Magnetic field observation as Voyager 1 entered the heliosheath depletion region. Science 12 (341), 147–150. https://doi.org/10.1126/science.1235451.
Camporeale, 2019. The challenge of machine learning in space weather: nowcasting and forecasting. Space Weather 17, 1166–1207. https://doi.org/10.1029/2018SW002061.
Cannon, P.S., et al., 2013. Extreme Space Weather: Impacts on Engineered Systems. Royal Academy of Engineering, London, U.K.
Carrington, R.C., 1860. Description of a singular appearance seen on the Sun on September 1, 1859. Mon. Not. R. Astron. Soc. 20, 13.
Campbell, W.H., 1980. Observation of electric currents in the Alaska oil pipeline resulting from auroral electrojet current sources. Geophys. J. R. Astr. Soc. 61, 437–449.
Chane, E., Saur, J., Neubauer, F.M., Raeder, J., Poedts, S., 2012. Observational evidence of Alfven wings at the Earth. J. Geophys. Res. 117, A09217. doi:10.1029/2012JA017628.
Chapman, S., Ferraro, V.C.A., 1932. A new theory of magnetic storms. Terr. Magn. Atmos. Electr. 37 (2), 147–156. doi:10.1029/TE037i002p00147.
Chapman, S., 1953. Polar and tropical aurorae: and the isoauroral diagram. Proc. Indian Acad. Sci. 37, 175–188.
Chaston, C.C., Bonnell, J.W., Wygant, J.R., Reeves, G.D., Baker, D.N., Melrose, D.B., Cairns, I.H., 2017. Radial transport of radiation belt electrons in kinetic field-line resonances. Geophys. Res. Lett. 44 (16), 8140–8148. https://doi.org/10.1002/2017GL074587.

Chou, K., Wang, A., Yu, W., Wang, J., 2019. Laboratory experiments on dusty spacesuit charging and arcing in plasma. IEEE Trans. Plasma Sci. 47, 3898–3904. doi:10.1109/tps.2019.2922243.

Cid, C., Palacios, J., Saiz, E., Guerrero, A., Cerrato, Y., 2014. On extreme geomagnetic storms. J. Space Weather Space Clim. 4, A28. doi:10.1051/swsc/2014026.

Cid, C., Saiz, E., Guerrero, A., Palacios, J., Cerrato, Y., 2015. A Carrington-like geomagnetic storm observed in the 21st century. J. Space Weather Space Clim. 5, A16. doi:10.1051/swsc/201517.

Cliver, E.W., Feynman, J., Garret, H.B., 1990. An estimate of the maximum speed of the solar wind, 1938-1989. J. Geophys. Res. 95 (A10), 17103–17112.

Cliver, E.W., Svalgaard, L., 2004. The 1859 solar-terrestrial disturbance and the current limits of extreme space weather activity. Sol. Phy. 224, 407–422.

Cliver, E.W., Dietrich, W.F., 2013. The 1859 space weather event revisited: limits of extreme activity. J. Space Weather Space Clim. 3, A31. doi:10.1051/swsc/201353.

Cliver, E.W., Mekhaldi, F., Muscheler, R., 2020b. Solar Longitude Distribution of High-energy Proton Flares: Fluences and Spectra. Astrophys. J. Lett. 900 (1), L11.

Connerney, J.E.P., Acuna, M.H., Wasilewski, P.J., Kletetschka, G., Ness, N.F., Reme, H., Lin, R.P., Mitchell, D.L., 2001. The global magnetic field of Mars and implications for crustal evolution. Geophys. Res. Lett. 28 (21), 4015–4018.

Cooper, A., et al., 2021. A global environmental crisis 42,000 years ago. Science 371, 811–818.

Covington, A.E., 1947. Microwave solar noise observations during the eclipse of November 23, 1946. Nature 159, 405–406.

Crosby, N.B., 2007. Major radiation environments in the heliosphere and their implications for interplanetary travel. In: Bothmer, V., Daglis, I.A. (Eds.), Space Weather Physics and Effects, pp. 131–165.

Crutzen, P.J., 1970. The influence of nitrogen oxides on the atmospheric ozone content. Quart. J. R. Met. Soc. 96, 320–325.

Dachev, T.P., Bankov, N.G., Tomov, B.T., Matviichuk, Y.N., Dimitrov, P.G., Hader, D.-P., Horneck, G., 2017. Overview of the ISS radiation environment observed during the ESA EXPOSE-R2 mission in 2014?2016. Space Weather 15, 1475–1489. https://doi.org/10.1002/2016SW001580.

Desai, M., Giacalone, J., 2016. Large gradual solar energetic particles events. Living Rev. Sol. Phys. 13, 3. doi:10.1007/s41116-016-0002-5.

Dessler, A.D., Parker, E.N., 1959. Hydromagnetic theory of geomagnetic storms. J. Geophys. Res. 64 (12), 2239–2252.

Dialynas, K., Krimigis, S.M., Decker, R.B., Mitchell, D.G., 2019. Plasma pressures in the heliosheath from Cassini ENA and Voyager 2 measurements: validation by the Voyager 2 heliopause crossing. Geophys. Res. Lett. 46, 7911–7919. https://doi.org/10.1029/2019GL083924.

Dmitriev, A.V., Suvorova, A.V., Chao, J.-K., Wang, C.B., Rastaetter, L., Panasyuk, M.I., Lazutin, L.L., Kovtyukh, A.S., Veselovsky, I.S., Myagkova, I.N., 2014. Anomalous dynamics of the extremely compressed magnetosphere during 21 January 2005 magnetic storm. J. Geophys. Res. Space Physics 119, 877–896. doi:10.1002/2013JA019534.

Drury, L.O., Voelk, J.H., Aug. 15, 1981. Hydromagnetic shock structure in the presence of cosmic rays. Astrophys. J. Part 1, 248, 344–351.

Dryer, M., Smith, Z., Fry, C.D., Sun, W., Deehr, C.S., Akasofu, S.-I., 2004. Real-time shock arrival predictions during the "Halloween 2003 epoch". Space Weather 2, S09001. doi:10.1029/2004SW000087.

Duggal, S.P., 1979. Relativistic solar cosmic rays. Rev. Geophys. 17 (5), 1021–1058. doi:10.1029/RG017i005p01021.

Dyer, C., Lei, F., 2001. Monte Carlo calculations of the influence on aircraft radiation environments of structures and solar particle events. IEEE Trans. Nucl. Sci. 48 (6), 1987–1995. doi:10.1109/23.983161.

Eastwood, J.P., Biffis, E., Hapgood, M.A., Green, L., Bisi, M.M., Bentley, R.D., Wicks, R., McKinnell, L.-A., Gibbs, M., Burnett, C., 2017. The economic impact of space weather: where do we stand? Risk Anal. 37, 206–218. https://doi.org/10.1111/risa.12765.

Ebihara, Y., Hayakawa, H., Iwahashi, K., Tamazawa, H., Kawamura, A.D., Isobe, H., 2017. Possible cause of extremely bright aurora witnessed in East Asia on 17 September 1770. Space Weather 15, 1373–1382. doi:10.1002/2017SW001693.

Ebihara, Y., 2019. Simulation study of near-Earth space disturbances: 1. magnetic storms. Prog. Earth Planet Sci. 6, 16. https://doi.org/10.1186/s40645-019-0264-3.

Ebihara, Y., Tanaka, T., 2015. Substorm simulation: formation of westward traveling surge. J. Geophys. Res. Space Physics 120, 10466–10484. doi:10.1002/2015JA021697.

Ebihara, Y., Tanaka, T., 2021. How do auroral substorms depend on Earth's dipole magnetic moment? J. Geophys. Res. Space Phys. 126, e2020JA028009, https://doi.org/10.1029/2020JA028009.

Ebihara, Y., Watari, S., Kumar, S., 2021. Prediction of geomagnetically induced currents (GICs) flowing in Japanese power grid for Carrington-class magnetic storms. Earth Planets Space 73, 163. https://doi.org/10.1186/s40623-021-01493-2.

Ebihara, Y., Fok, M.-C., Sazykin, S., Thomsen, M.F., Hairston, M.R., Evans, D.S., Rich, F.J., Ejiri, M., 2005. Ring current and the magnetosphere-ionosphere coupling during the superstorm of 20 November 2003. J. Geophys. Res. 110, A09S22. doi:10.1029/2004JA010924.

Ebihara, Y., Nishitani, N., Kikuchi, T., Ogawa, T., Hosokawa, K., Fok, M.-C., Thomsen, M.F., 2009. Dynamical property of storm time subauroral rapid flows as a manifestation of complex structures of the plasma pressure in the inner magnetosphere. J. Geophys. Res. 114, A01306. doi:10.1029/2008JA013614.

Egan, H., Jarvinen, R., Ma, Y., Brain, D., 2019. Planetary magnetic field control of ion escape from weakly magnetized planets. Mon. Not. R. Astron. Soc. 488, 2108–2120.

Ehresmann, B., Hassler, D.M., Zeitlin, C., Guo, J., Wimmer-Schweingruber, R.F., Matthia, D., et al., 2018. Energetic particle radiation environment observed by RAD on the surface of Mars during the September 2017 event. Geophys. Res. Lett. 45, 5305−5311. https://doi.org/10.1029/2018GL077801.

Elkington, S.R., Hudson, M.K., Chan, A.A., 1999. Acceleration of relativistic electrons via drift-resonant interaction with toroidal-mode Pc-5 ULF oscillations. Geophys. Res. Lett. 26 (21), 3273–3276. doi:10.1029/1999GL003659.

Emslie, A.G., Dennis, B.R., Shih, A.Y., Chamberlin, P.C., Mewaldt, R.A., Moore, C.S., Share, G.H., Vourlidas, V., Welsch, B.T., 2012. Global energetics of thirty-eight large solar eruptive events. Astrophys. J. 759, 71.

Engineering R.A., 2013. Extreme Space Weather: Impacts on Engineered Systems and Infrastructure.

Erinmez, I.A., Kappenman, J.G., Radasky, W.A., 2002. Management of the geomagnetically induced current risks on the national grid company's electric power transmission system. J. Atmos. Sol. Terr. Phys. 64 (5-6), 743–756. doi:10.1016/S1364-6826(02)00036-6.

Facius, R., Reitz, G., 2007. Space weather impacts on space radiation protection. In: Bothmer, V., Daglis, I.A. (Eds.), Space Weather Physics and Effects, Springer Praxis Books, pp. 289–345.

Fairfield, D.H., Iver, H.C., Desch, M.D., Szabo, A., Lazarus, A.J., Aellig, M.R., 2001. The location of low Mach number bow shocks at Earth. J. Geophys. Res. 106 (A11), 25361–25376. doi:10.1029/2000JA000252.

Fairfield, D.H., Cahill, L.J., 1966. Transition region magnetic field and polar magnetic disturbances. J. Geophys. Res. 71 (1), 155–169. doi:10.1029/JZ071i001p00155.

Fernberg, P., 2012. One-Dimensional Earth Resistivity Models for Selected Areas of Continental United States and Alaska. EPRI Technical Update 1026430, Palo Alto, CA.

Fields, B.D., Athanssiadou, T., Johnson, S.R., 2008. Supernova collisions with the heliosphere. Astron. J. 678, 549–562.

Fok, M.-C., Wolf, R.A., Spiro, R.W., Moore, T.E., 2001. Comprehensive computational model of Earth's ring current. J. Geophys. Res. 106 (A5), 8417–8424. doi:10.1029/2000JA000235.

Fok, M.-C., Kang, S.-B., Ferradas, C.P., Buzulukova, N.Y., Glocer, A., Komar, C.M., 2021. New developments in the comprehensive inner magnetosphere-ionosphere model. J. Geophys. Res. Space Phys. 126, e2020JA028987. https://doi.org/10.1029/2020JA028987.

Forbush, S.E., 1946. Three unusual cosmic-ray increases possibly due to charged particles from the Sun. Phys. Rev. 70, 771–772.

Forget, F., Wordsworth, R., Millour, E., Madeleine, J.-B., Kerber, L., Leconte, J., Marcq, E., Haberle, R.M., 2013. 3D modelling of the early martian climate under a denser CO2 atmosphere: temperatures and CO2 ice clouds. Icarus 222, 81–99.

Foster, J.C., Buonsanto, M.J., Mendillo, M., Nottingham, D., Rich, F.J., Denig, W., 1994. Coordinated stable auroral red arc observations: Relationship to plasma convection. J. Geophys. Res. 99 (A6), 11429–11439. doi:10.1029/93JA03140.

Foster, J.C., Burke, W.J., 2002. SAPS: a new characterization for sub-auroral electric fields. EOS 83, 393–394.

Foster, J.C., Vo, H.B., 2002. Average characteristics and activity dependence of the subauroral polarization stream. J. Geophys. Res. 107 (A12), 1475. doi:10.1029/2002JA009409.

Frahm, R.A., Winningham, J.D., Sharber, J.R., Link, R., Crowley, G., Gaines, E.E., Chenette, D.L., Anderson, B.J., Potemra, T.A., 1997. The diffuse aurora: a significant source of ionization in the middle atmosphere. J. Geophys. Res. 102 (D23), 28203–28214. doi: 10.1029/97JD02430.

Freiherr von Forstner, J.L., Guo, J., Wimmer-Schweingruber, R.F., et al., 2018. Using Forbush decreases to derive the transit time of ICMEs propagating from 1 AU to Mars. J. Geophys. Res. Space Phys. 123, 39–56. https://doi.org/10.1002/2017JA024700.

Fry, C.D., Sun, W., Deehr, C.S., Dryer, M., Smith, Z., Akasofu, S.-I., Tokumaru, M., Kojima, M., 2001. Improvements to the HAF solar wind model for space weather predictions. J. Geophys. Res. 106 (A10), 20985–21001. doi:10.1029/2000JA000220.

Fujita, S., Tanaka, T., Kikuchi, T., Fujimoto, K., Hosokawa, K., Itonaga, M., 2003a. A numerical simulation of the geomagnetic sudden commencement: 1. Generation of the field-aligned current associated with the preliminary impulse. J. Geophys. Res. 108 (A12), 1416. doi:10.1029/2002JA009407.

Fujii, I., Ookawa, T., Nagamachi, S., et al., 2015. The characteristics of geoelectric fields at Kakioka, Kanoya, and Memambetsu inferred from voltage measurements during 2000 to 2011. Earth Planet Sp **67**, 62. https://doi.org/10.1186/s40623-015-0241-z.

Fujita, S., Tanaka, T., Kikuchi, T., Fujimoto, K., Itonaga, M., 2003b. A numerical simulation of the geomagnetic sudden commencement: 2. Plasma processes in the main impulse. J. Geophys. Res. 108 (A12), 1417. doi:10.1029/2002JA009763.

Fujita, S., Kataka, R., Fujii, I., Pulkkinen, A., Watari, S., 2016. Extremely severe space weather and geomagnetically induced currents in regions with locally heterogeneous ground resistivity. Earth Planets Space 68, 51. doi:10.1086/s40623-016-0428-y.

Fukuda, Y., Kataoka, R., Uchida, H.A., Miyoshi, Y., Hampton, D., Shiokawa, K., Ebihara, Y., Whiter, D., Iwagami, N., Seki, K., 2017. First evidence of patchy flickering aurora modulated by multi-ion electromagnetic ion cyclotron waves. Geophys. Res. Lett. 44, 3963–3970. doi:10.1002/2017GL072956.

Fukushima, N., 1969. Equivalence in ground geomagnetic effect of Chapman-Vestine's and Birkeland-Alfven's current systems for polar magnetic storms. Rep. Ionos.Space Res. Jap 23, 219–227.

Fukushima, N., 1976. Generalized theorem for no ground magnetic effect of vertical currents connected with Pedersen currents in the uniform conducting ionosphere. Rep. Ionos. Space Res. 30, 35–50.
Ganse, U., Kilian, P., Vainio, R., Spanier, F., 2012. Emission of type II radio bursts - single-beam versus two-beam scenario. Solar Phys. 280, 551–560. doi:10.1007/s11207-012-0077-7.
Garraffo, C., Drake, J.J., Cohen, O., 2016. The space weather of Proxima Centauri b. Astrophys. J. Lett. 833 (1), L4 Number.
Gerard, J.-C., Soret, L., Shematovich, V.I., Bisikalo, D.V., Bougher, S.W., 2017. The Mars diffuse aurora: A model of ultraviolet and visible emissions. Icarus 288, 284–294. https://doi.org/10.1016/j.icarus.2017.01.037.
Girgis, K.M., Hada, T., Matsukiyo, S., 2022. Seasonal variation and geomagnetic storm index effects on the proton flux response in the South Atlantic Anomaly by test particle simulations. J. Atmos. Sol.-Terr. Phys. 228, 105808, https://doi.org/10.1016/j.jastp.2021.105808.
Glassmeier, K.-H., Vogt, J., 2010. Magnetic polarity transitions and biospheric effects: historical perspective and current development. Space Sci. Rev. 155, 387–410. doi:10.1007/s11214-010-9659-6.
Gleeson, L.J., Axford, W.I., 1968. Modulation of galactic cosmic rays. Astrophys. J. 154, 1011–1026.
Gold, T., 1955. Discussion of shock waves and rarefied gases. In: van de Hulst, JC, Burgers, JM (Eds.), Gas Dynamics of Cosmic Clouds. North-Holland Publishing Co., Amsterdam, pp. 103.
Gold, T., Palmer, D.R., 1956. The solar outburst, 23 February 1956 - Observations by the Royal Greenwich Observatory. J. Atmos. Terr. Phys. 8, 287–290.
Goldstein, J., Burch, J.L., Sandel, B.R., 2005. Magnetospheric model of subauroral polarization stream. J. Geophys. Res. 110, A09222. doi:10.1029/2005JA011135.
Gopalswamy, N., Lara, A., Yashiro, S., Kaiser, M.L., Howard, R.A., 2001. Predicting the 1-AU arrival times of coronal mass ejections. J. Geophys. Res. 106 (A12), 29207–29217. doi: 10.1029/2001JA000177.
Gopalswamy, N., Yashiro, S., Krucker, S., Stenborg, G., Howard, R.A., 2004. Intensity variation of large solar energetic particle events associated with coronal mass ejections. J. Geophys. Res. 109, A12105. doi:10.1029/2004JA010602.
Gopalswamy, N., Yashiro, S., Liu, Y., Michalek, G., Vourlidas, A., Kaiser, M.L., Howard, R.A., 2005. Coronal mass ejections and other extreme characteristics of the 2003 October-November solar eruptions. J. Geophys. Res. 110, A09S15. doi:10.1029/2004JA010958.
Gopalswamy, N., Xie, H., Yashiro, S., Akiyama, S., Makela, P., Usoskin, I.G., 2012. Properties of ground level enhancement events and the associated solar eruptions during solar cycle 23. Space Sci. Rev. 171, 23–60.
Gopalswamy, N., 2016. History and development of coronal mass ejections as a key player in solar terrestrial relationship. Geosci. Lett. 3, 8.
Gopalswamy, N., 2018. Extreme solar eruptions and their space weather consequences. In: Buzulukova, N. (Ed.), Extreme Events in Geospace, Elsevier, pp. 37–63.
Gosling, J.T., 1993. The solar flare myth. J. Geophys. Res. 98 (A11), 18937–18949.
Green, J.C., Kivelson, M.G., 2004. Relativistic electrons in the outer radiation belt: differentiating between acceleration mechanisms. J. Geophys. Res. 109, A03213. doi:10.1029/2003JA010153.
Green, J.L., Boardsen, S., 2006. Duration and extent of the great auroral storm of 1859. Adv. Space Res. 38, 130–135.
Greenwald, R.A., 2021. History of the Super Dual Auroral Radar Network (SuperDARN)-I: pre-SuperDARN developments in high frequency radar technology for ionospheric research and selected scientific results. Hist. Geo Space. Sci. 12, 77–93. https://doi.org/10.5194/hgss-12-77-2021.

Grenfell, J.L., Lehmann, R., Mieth, P., Langematz, U., Steil, B., 2006. Chemical reaction pathways affecting stratospheric and mesospheric ozone. J. Geophys. Res. 111, D17311. doi:10.1029/2004JD005713.
Gronoff, G., Arras, P., Baraka, S., Bell, J.M., Cessateur, G., Cohen, O., et al., 2020. Atmospheric escape processes and planetary atmospheric evolution. J. Geophys. Res. Space Phys. 125, e2019JA027639. https://doi.org/10.1029/2019JA027639.
Guillon, S., Toner, P., Gibson, L., Boteler, D., 2016. A colorful blackout. IEEE Power Energ. Mag. 14 (6), 59–71.
Guo, J., et al., 2018. Modeling the evolution and propagation of 10 September 2017 CMEs and SEPs arriving at Mars constrained by remote sensing and in situ measurement. Space Weather 16, 1156–1169.
Guo, J., Wimmer-Schweingruber, R.F., Grande, M., Lee-Payne, Z.H., Matthia, D., 2019. Ready functions for calculating the Martian radiation environment. J. Space Weather Space Clim. 9, A7.
Haider, S.A., Masoom, J., 2019. Modeling of Diffuse Aurora due to Precipitation of H+-H and SEP Electrons in the Nighttime Atmosphere of Mars: Monte Carlo Simulation and MAVEN Observation. J. Geophys. Res. Space Phys. 124, 9566–9576. https://doi.org/10.1029/2019JA026688.
Hakamada, K., Akasofu, S.I., 1982. Simulation of three-dimensional solar wind disturbances and resulting geomagnetic storms. Space Sci. Rev. 31, 3–70. https://doi.org/10.1007/BF00349000.
Halekas, J.S., Delory, G.T., Brain, D.A., Lin, R.P., Fillingim, M.O., Lee, C.O., Mewaldt, R.A., Stubbs, T.J., Farrell, W.M., Hudson, M.K., 2007. Extreme lunar surface charging during solar energetic particle events. Geophys. Res. Lett. 34, L02111. doi:10.1029/2006GL028517.
Halekas, J.S., Delory, G.T., Lin, R.P., Stubbs, T.J., Farrell, W.M., 2009. Lunar surface charging during solar energetic particle events: Measurement and prediction. J. Geophys. Res. 114, A05110. doi:10.1029/2009JA014113.
Hapgood, M.A., 2011. Towards a scientific understanding of the risk from extreme space weather. Adv. Space Res. 47, 2059–2072.
Hapgood, M., 2012. Preparing for the coming space weather storm. Nature 484, 311–313.
Hapgood, M., et al. (2016), Summary of space weather worst-case environments. Revised edition, Technical Report, RAL-TR-2016-006.
Hapgood, M., 2019. The great storm on May 1921: an exemplar of a dangerous space weather event. Space Weather 17, 950–975. doi:10.1029/2019SW002195.
Harnett, E.M., Cash, M., Winglee, R.M., 2013. Substorm and storm time ionospheric particle flux at the Moon while in the terrestrial magnetosphere. Icarus 224 (1), 218–227. https://doi.org/10.1016/j.icarus.2013.02.022.
Harrison, R.A., Bryans, P., Simnett, G.M., Lyons, M., 2003. Coronal dimming and the coronal mass ejection onset. A&A 400, 1071–1083.
Hassler, D.M., et al., 2014. Mars' surface radiation environment measured with the Mars Science Laboratory's Curiosity Rover. Science 343, 1244797-1-6.
Hassler, D.M., et al., 2018. Space weather on the surface of Mars: impact of the September 2017 events. Space Weather 16, 1702–1708. https://doi.org/10.1029/2018SW001959.
Hayakawa, H., Iwahashi, K., Tamazawa, H., Isobe, H., Kataoka, R., Ebihara, Y., Miyahara, H., Kawamura, A.D., Shibata, K., 2016. East Asian observations of low-latitude aurora during the Carrington magnetic storm. Publ. Astron. Soc. Aust. 68 (6), 99. https://doi.org/10.1093/pasj/psw097.
Hill, T.W., Dessler, A.J., Wolf, R.A., 1976. Mercury and Mars: the role of ionospheric conductivity in the acceleration of magnetospheric particles. Geophys. Res. Lett. 3, 429–432. doi:10.1029/GL003i008p00429.
Hirooka, K., 1971. Archaeomagnetic study for the past 2000 years in southwest JapanMemoirs of the Faculty of Science, 2. Kyoto University. College of Science, Kyoto Imperial University, Series of Geol Mineral XXXVIII, pp. 167–207.

Hoffman, R.A., Cahill Jr., L.J., Anderson, R.R., Maynard, N.C., Smith, P.H., Fritz, T.A., Williams, D.J., Konradi, A., Gurnett, D.A., 1975. Explorer 45 (S3-A) observations of the magnetosphere and magnetopause during the August 4-6, 1972. Magnet. Storm Period 80 (31), 4287–4296.

Hosokawa, K., Miyoshi, Y., Ozaki, M., et al., 2020. Multiple time-scale beats in aurora: precise orchestration via magnetospheric chorus waves. Sci. Rep. 10, 3380. https://doi.org/10.1038/s41598-020-59642-8.

Hotta, H., Rempel, M., Yokoyama, T., 2015. Efficient small-scale dynamo in the solar convection zone. Astrophys. J. 803, 42.

Hotta, H., Kusano, K., 2021. Solar differential rotation reproduced with high-resolution simulation. Nat. Astron. https://doi.org/10.1038/s41550-021-01459-0.

Howard, R.A., Michels, D.J., Sheeley Jr., N.R., Koomen, M.J., 1982. The observation of a coronal transient directed at earth. Astrophys. J. 263, L101–L104.

Huttunen, K.E.J., Kilpua, S.P., Pulkkinen, A., Viljanen, A., Tanskanen, E., 2008. Solar wind drivers of large geomagnetically induced currents during the solar cycle 23. Space Weather 6, S10002. doi:10.1029/2007SW000374.

Iijima, T., Potemra, T.A., 1978. Large-scale characteristics of field-aligned currents associated with substorms. J. Geophys. Res. 83 (A2), 599–615. doi:10.1029/JA083iA02p00599.

Inoue, S., Kusano, K., Buchner, J., et al., 2018. Formation and dynamics of a solar eruptive flux tube. Nat. Commun. 9, 174. https://doi.org/10.1038/s41467-017-02616-8.

Ishii, M., Shiota, D., Tao, C., et al., 2021. Space weather benchmarks on Japanese society. Earth Planets Space 73, 108. https://doi.org/10.1186/s40623-021-01420-5.

Iwai, K., Shiota, D., Tokumaru, M., et al., 2021. Validation of coronal mass ejection arrival-time forecasts by magnetohydrodynamic simulations based on interplanetary scintillation observations. Earth Planets Space 73, 9. https://doi.org/10.1186/s40623-020-01345-5.

Jacobs, J.A., 1994. Reversals of the Earth's Magnetic Field. Cambridge University Press, Cambridge.

Jackman, C.H., DeLand, M.T., Labow, G.J., Fleming, E.L., Weisenstein, D.K., Ko, M.K.W., Sinnhuber, M., Russell, J.M., 2005. Neutral atmospheric influences of the solar proton events in October–November 2003. J. Geophys. Res. 110, A09S27. doi:10.1029/2004JA010888.

Jakovsky, B.M., Slipski, M., Benna, M., Mahaffy, P., Elrod, M., Yelle, R., Stone, S., Alsaeed, N., 2017. Mar's atmospheric history derived from upper-atmosphere measurements of 38Ar/36Ar. Science 355, 1408–1410.

Jaynes, A.N., Baker, D.N., Singer, H.J., Rodriguez, J.V., Lotoaniu, T.M., Ali, A.F., et al., 2015. Source and seed populations for relativistic electrons: their roles in radiation belt changes. J. Geophys. Res. Space Phys. 120 (9), 7240–7254. https://doi.org/10.1002/2015JA021234.

Jiggens, P., Chavy-Macdonald, M.-A., Santin, G., Menicucci, A., Evans, H., Hilgers, A., 2014. The magnitude and effects of extreme solar particle events. J. Space Weather Space Clim. 4, A20. doi:10.1051/swsc/2014017.

Jiggens, P., Clavie, C., Evans, H., O'Brien, T.P., Witasse, O., Mishev, A.L., et al., 2019. In situ data and effect correlation during September 2017 solar particle event. Space Weather 17, 99–117. https://doi.org/10.1029/2018SW001936.

Jin, M., Manchester, W.B., van der Holst, B., Sokolov, I., Toth, G., Mullinix, R.E., Taktakishvili, A., 2017. Data-constrained coronal mass ejections in a global magnetohydrodynamics model. Astrophys. J. 834, 173.

Jokipii, J., Kóta, J., 2000. Galactic and Anomalous Cosmic Rays in the Heliosphere. Astrophys. Space Sci. 274, 77–96. https://doi.org/10.1023/A:1026535603934.

Jordan, A.P., Stubbs, T.J., Wilson, J.K., Schwadron, N.A., Spence, H.E., 2015. Dielectric breakdown weathering of the Moon's polar regolith. J. Geophys. Res. Planets 120, 210–225. doi:10.1002/2014JE004710.

Jordan, A.P., Stubbs, T.J., Wilson, J.K., Schwadron, N.A., Spence, H.E., 2017. The rate of dielectric breakdown weathering of lunar regolith in permanently shadowed regions. Icarus 283, 352–358. doi:10.1016/j.icarus.2016.08.027.

Jorgensen, J.L., Benn, M., Connerney, J.E.P., Denver, T., Jorgensen, P.S., Andersen, A.C., Bolton, S.J., 2021. Distribution of interplanetary dust detected by the Juno spacecraft and its contribution to the Zodiacal Light. J. Geophys. Res. Planets 126, e2020JE006509. https://doi.org/10.1029/2020JE006509.

Justus, C.G., Duvall, A., Keller, V.W., 2006. Validation of Mars Global Reference Atmospheric Model (Mars-GRAM 2001) and planned new features. Adv. Space Res. 38, 2633–2638. doi:10.1016/j.asr.2006.07.007.

Justus, C.G., and D.L. Johnson, 2001. Mars Global Reference Atmospheric Model 2001 Version (Mars-GRAM 2001), NASA TM-2001-210961.

Kaneko, T., Yokoyama, T., 2017. Reconnection Condensation Model for Solar Prominence Formation. Astrophys. J. 845 (1), https://doi.org/10.3847/1538-4357/aa7d59.

Kappenman, J.G., 2003. Storm sudden commencement events and the associated geomagnetically induced current risks to ground-based systems at low-latitude and midlatitude locations. Space Weather 1 (3), 1016. doi:10.1029/2003SW000009.

Karsberg, A., Swedenborg, G., Wyke, K., 1959. The influences of earth magnetic currents on telecommunication lines, Tele (English ed.). Televerket (Swedish Telecom), Stockholm, pp. 1–21.

Kasahara, S., Miyoshi, Y., Yokota, S., Mitani, T., Kasahara, Y., Matsuda, S., Kumamoto, A., Matsuoka, A., Kazama, Y., Frey, H.U., Anvelopoulos, V., Kurita, S., Keika, K., Seki, K., Shinohara, I., 2018. Pulsating aurora from electron scattering by chorus waves. Nature 554, 337–340. doi:10.1038/nature25505.

Kataoka, R., Fukunishi, H., Fujita, S., Tanaka, T., Itonaga, M., 2004. Transient response of the Earth's magnetosphere to a localized density pulse in the solar wind: simulation of traveling convection vortices. J. Geophys. Res. 109, A03204. doi:10.1029/2003JA010287.

Kataoka, R., Nakano, S., 2021. Reconstructing solar wind profiles associated with extreme magnetic storms: A machine learning approach. Geophys. Res. Lett. 48. e2021GL096275. https://doi.org/10.1029/2021GL096275.

Kataoka, R., Fairfield, D.H., Sibeck, D.G., Rastaetter, L., Fok, M.-C., Nagatsuma, T., Ebihara, Y., 2005b. Magnetosheath variations during the storm main phase on 20 November 2003: evidence for solar wind density control of energy transfer to the magnetosphere. Geophys. Res. Lett. 32, L21108. doi:10.1029/2005GL024495.

Kataoka, R., Watari, S., Shimada, N., Shimazu, H., Marubashi, K., 2005a. Downstream structures of interplanetary fast shocks associated with coronal mass ejections. Geophys. Res. Lett. 32, L12103. doi:10.1029/2005GL022777.

Kataoka, R., Miyoshi, Y., 2006. Flux enhancement of radiation belt electrons during geomagnetic storms driven by coronal mass ejections and corotating interaction regions. Space Weather 4, S09004. doi:10.1029/2005SW000211.

Kataoka, R., Nishitani, N., Ebihara, Y., Hosokawa, K., Ogawa, T., Kikuchi, T., Miyoshi, Y., 2007. Dynamic variations of a convection flow reversal in the subauroral postmidnight sector as seen by the SuperDARN Hokkaido HF radar. Geophys. Res. Lett. 34, L21105. doi:10.1029/2007GL031552.

Kataoka, R., Miyoshi, Y., 2008. Magnetosphere inflation during the recovery phase of geomagnetic storms as an excellent magnetic confinement of killer electrons. Geophys. Res. Lett. 35, L06S09. doi:10.1029/2007GL031842.

Kataoka, R., Pulkkinen, A., 2008. Geomagnetically induced currents during intense storms driven by coronal mass ejections and corotating interacting regions. J. Geophys. Res. 133, A03S12. doi:10.1029/2007JA012487.

Kataoka, R., Hosokawa, K., Nishitani, N., Miyoshi, Y., 2009. SuperDARN Hokkaido radar observation of westward flow enhancement in subauroral latitudes. Ann. Geophys. 27, 1695–1699.

Kataoka, R., Ebisuzaki, T., Kusano, K., Shiota, D., Inoue, S., Yamamoto, T., Tokumaru, M., 2009. Three-dimensional magnetohydrodynamic (MHD) modeling of the solar wind

structures associated with 13 December 2006 coronal mass ejection. J. Geophys. Res. 114, A10102. doi:10.1029/2009JA014167.
Kataoka, R., Miyoshi, Y., Hampton, D., Ishii, T., Kozako, H., 2012b. Pulsating aurora beyond the ultra-low frequency range. J. Geophys. Res. 117, A08336. doi:10.1029/2012 JA017987.
Kataoka, R., Ebisuzaki, T., Miyahara, H., Maruyama, S., 2013a. Snowball Earth events driven by starbursts of the Milky Way Galaxy. New Astron. 21, 50–62. doi:10.1016/j.newast.2012.11.005.
Kataoka, R., Miyoshi, Y., Shigematsu, K., Hampton, D., Mori, Y., Kubo, T., Yamashita, A., Tanaka, M., Takahei, T., Nakai, T., Miyahara, H., Shiokawa, K., 2013b. Stereoscopic determination of all-sky altitude map of aurora using two ground-based Nikon DSLR cameras. Ann. Geophys. 31, 1543–1548. https://doi.org/10.5194/angeo-31-1543-2013 2013.
Kataoka, R., 2013. Probability of occurrence of extreme magnetic storms. Space Weather 11, 214–218. doi:10.1002/swe.20044.
Kataoka, R., Shiota, D., Kilpua, E., Keika, K., 2015a. Pileup accident hypothesis of magnetic storm on 2015 March 17. Geophys. Res. Lett. 42, 5155–5161. doi:10.1002/2015 GL064816.
Kataoka, R., Ebisuzaki, T., Miyahara, H., Nimura, T., Tomida, T., Sato, T., Maruyama, S., 2014. The Nebula winter: the united view of the snowball Earth, mass extinctions, and explosive evolution in the late Neoproterozoic and Cambrian periods. Gondwana Res. 25 (3), 1153–1163. doi:10.1016/j.gr.2013.05.003.
Kataoka, R., Ngwira, C., 2016. Extreme geomagnetically induced currents. Progr. Earth Planet. Sci. 3, 23. doi:10.1186/s40645-016-0101-x.
Kataoka, R., Asaoka, Y., Torii, S., Terasawa, T., Ozawa, S., Tamura, T., Shimizu, Y., Akaike, Y., Mori, M., 2016. Relativistic electron precipitation at International Space Station: space weather monitoring by Calorimetric Electron Telescope. Geophys. Res. Lett. 43, 4119–4125. doi:10.1002/2016GL068930.
Kataoka, R., Isobe H., Hayakawa H., Tamazawa H., Kawamura A.D., Miyahara H., Iwasaki K., Yamamoto K., Takei M., Terashima T., Suzuki H., Fujiwara Y., Nakamura T., 2017. Historical space weather monitoring of prolonged aurora activities in Japan and in China, 15 (2), 392–402, doi:10.1002/2016SW001493.
Kataoka, R., Iwahashi, K., 2017. Inclined zenith aurora over Kyoto on 17 September 1770: Graphical evidence of extreme magnetic storm. Space Weather 15, 1314–1320. doi: 10.1002/2017SW001690.
Kataoka, R., Sato, T., Miyake, S., Shiota, D., Kubo, Y., 2018. Radiation dose Nowcast for the ground level enhancement on 10-11 September 2017. Space Weather 16. https://doi.org/10.1029/2018SW001874.
Kataoka, R., Uchino S., Fujiwara Y., Fujita S., Yamamoto K., 2019. Fan-shaped aurora as seen from Japan during a great magnetic storm on 11 February 1958, 9, A16, https://doi.org/10.1051/swsc/2019013.
Kataoka, R., Kazama, S., 2019. A watercolor painting of northern lights seen above Japan on 11 February 1958. J. Space Weather Space Clim. 9, A28. https://doi.org/10.1051/swsc/2019027.
Kataoka, R., 2020. Extreme geomagnetic activities: a statistical study, 72, 124. https://doi.org/10.1186/s40623-020-01261-8.
Kataoka, R., Yamamoto, K., Fujiwara, Y., Shiomi, K., Kokubun, N., 2020b. Pheasant Tail: Consideration of the shape of the red sign in the Nihon-Shoki. SOKENDAI Rev. Cult. Soc. Stud. 16, 17–28.
Kataoka, R., Chaston, C.C., Knudsen, D., Lynch, K.A., Lysak, R., Song, Y., Rankin, R., Murase, K., Sakanoi, T., Semeter, J., Watanabe, T.-H., Whiter, D., 2021. Small-scale dynamic aurora. Space Sci. Rev. 217, 17. https://doi.org/10.1007/s11214-021-00796-w.

Kataoka, R., Nakano, S., 2021. Auroral zone over the last 3000 years. J. Space Weather Space Clim. 11, 46. https://doi.org/10.1051/swsc/2021030.

Kataoka, R., Nakagawa, Y., Sato, T., 2015b. Radiation dose of aircrews during a solar proton event without ground-level enhancement. Ann. Geophys. 33, 75–78. doi:10.5194/angeo-33-75-2015.

Kataoka, R., Shiota, D., Fujiwara, H., Jin, H., Tao, C., Shinagawa, H., Miyoshi, Y., 2022. Unexpected space weather causing the reentry of 38 Starlink satellites in February 2022. J. Space Weather and Space Clim. https://doi.org/10.31223/X5GH0X.

Katsavrias, C., Sandberg, I., Li, W., Podladchikova, O., Daglis, I.A., Papadimitriou, C., et al., 2019. Highly relativistic electron flux enhancement during the weak geomagnetic storm of April-May 2017. J. Geophys. Res. Space Phys. 124, 4402–4413. https://doi.org/10.1029/2019JA026743.

Kato, C., Kihara, W., Ko, Y., Kadokura, A., Kataoka, R., Evenson, P., Uchida, S., Kaimi, S., Nakamura, Y., Uchida, H.A., Murase, K., Munakata, K., 2021. New cosmic ray observations at Syowa Station in the Antarctic for space weather study. J. Space Weather Space Clim. 11, 31. https://doi.org/10.1051/swsc/2021005.

Keika, K., Ebihara, Y., Kataoka, R., 2015. What caused the rapid recovery of the Carrington storm? Earth Planets Space 67, 65. doi:10.1186/s40623-015-0234-y.

Kennel, C.F., Petschek, H.E., 1966. Limit on stably trapped particle fluxes. J. Geophys. Res. 71 (1), 1–28. doi:10.1029/JZ071i001p00001.

Kessler, D.J., et al., 2010. The Kessler syndrome: Implications to future space operations. Adv. Astron. Sci. 137, 47–62.

Kihara, W., Munakata, K., Kato, C., Kataoka, R., Kadokura, A., Miyake, S., et al., 2021. A peculiar ICME event in August 2018 observed with the Global Muon Detector Network. Space Weather 19, e2020SW002531. https://doi.org/10.1029/2020SW002531.

Kikuchi, T., Araki, T., Maeda, H., Maekawa, K., 1978. Transmission of polar electric fields to the Equator. Nature 273, 650–651.

Kikuchi, T., Chum, J., Tomizawa, I., et al., 2021. Penetration of the electric fields of the geomagnetic sudden commencement over the globe as observed with the HF Doppler sounders and magnetometers. Earth Planets Space 73, 10. https://doi.org/10.1186/s40623-020-01350-8.

Kilpua, E.K.J., Olspert, N., Grigorievskiy, A., Kapyla, M.J., Tanskanen, E., Miyahara, H., Kataoka, R., Pelt, J., Liu, Y.D., 2015a. Statistical study of strong and extreme geomagnetic disturbances and solar cycle characteristics. Astrophys. J. 806, 272. doi:10.1088/0004-637X/806/2/272.

Kilpua, E.K.J., Hietala, H., Turner, D.L., Koskinen, H.E.J., Pulkkinen, T.I., Rodriguez, J.V., Reeves, G.D., Claudepierre, S.G., Spence, H.E., 2015b. Unraveling the drivers of the storm time radiation belt response. Geophys. Res. Lett. 42, 3076–3084. doi:10.1002/2015GL063542.

Kilpua, E., Koskinen, H.E.J., Pulkkinen, T.I., 2017. Coronal mass ejections and their sheath regions in interplanetary space. Living Rev. Sol. Phys. 14, 5. https://doi.org/10.1007/s41116-017-0009-6.

Kilpua, E.K.J., Lugaz, N., Mays, M.L., Temmer, M., 2019. Forecasting the structure and orientation of earthbound coronal mass ejections. Space Weather 17, 498–526. https://doi.org/10.1029/2018SW001944.

Kimball, D.S., 1960. A study of the aurora of 1859Scientific Report No. 6. Geophysical Institute of the University of Fairbanks, Alaska, UAG-R109.

Kivelson, M.G., et al., 1997. The magetnic field and magnetosphere of Ganymede. Geophys. Res. Lett. 24 (17), 2155–2158.

Knipp, D.J., et al., 2016. The May 1967 great storm and radio disruption event: extreme space weather and extraordinary responses. Space Weather 14, 614–633. doi:10.1002/2016SW001423.

Knipp, D.J., Bernstein, V., Wahl, K., Hayakawa, H., 2021. Timelines as a tool for learning about space weather storms. J. Space Weather Space Clim. 11, 29. https://doi.org/10.1051/swsc/2021011.
Korte, M., Constable, C., 2011. Improving geomagnetic field reconstructions for 0?3 ka. 188, 3-4, 247-259. https://doi.org/10.1016/j.pepi.2011.06.017.
Kota, J., Jokipii, J.R., 2001. 3-D modeling of cosmic-ray transport in the heliosphere: toward solar maximum. Adv. Space Res. 27 (3), 529–534 2001.
Kozarev, K., Schwadron, N.A., Dayeh, M.A., Townsend, L.W., Desai, M.I., PourArsalan, M., 2010. Modeling the 2003 Halloween events with EMMREM: energetic particles, radial gradients, and coupling to MHD. Space Weather 8, S00E08. doi:10.1029/2009SW000550.
Kozyra, J.U., Shelley, E.G., Comfort, R.H., Brace, L.H., Cravens, T.E., Nagy, A.F., 1987. The role of ring current O+ in the formation of stable auroral red arcs. J. Geophys. Res. 92 (A7), 7487–7502. doi:10.1029/JA092iA07p07487.
Krimigis, S.M., Decker, R.B., Roelof, E.C., Hill, M.E., Armstrong, T.P., Gloeckler, G., 2013. Search for the exit: Voyager 1 at heliosphere's border with the Galaxy. Science 341, 144–147. https://doi.org/10.1126/science.1234721.
Kubota, Y., Kataoka, R., Den, M., et al., 2015. Global MHD simulation of magnetospheric response of preliminary impulse to large and sudden enhancement of the solar wind dynamic pressure. Earth Planet Space 67, 94. https://doi.org/10.1186/s40623-015-0270-7.
Kurita, S., Miyoshi, Y., Kasahara, S., Yokota, S., Kasahara, Y., Matsuda, S., Kumamoto, A., Matsuoka, A., Shinohara, I., 2018. Deformation of electron pitch angle distributions caused by upper-band chorus observed by the Arase satellite. Geophys. Res. Lett. 45. doi:10.1029/2018GL079104.
Kusano, K., Iju, T., Bamba, Y., Inoue, S., 2020. A physics-based method that can predict imminent large solar flares. Science 369, 587–591.
Kusano, K., Ichimoto, K., Ishii, M., et al., 2021. PSTEP: project for solar?terrestrial environment prediction. Earth Planets Space 73, 159. https://doi.org/10.1186/s40623-021-01486-1.
Kuwabara, T., et al., 2006. Real-time cosmic ray monitoring system for space weather. Space Weather 4, S08001. doi:10.1029/2005SW000204.
Lai, S.T., et al., 2018. Deep dielectric charging and spacecraft anomalies. In: Buzulukova, N. (Ed.), Extreme Events in Geospace, Elsevier, pp. 419–432.
Lakka, A., Pulkkinen, T.I., Dimmock, A.P., Myllys, M., Honkonen, I., Palmroth, M., 2018. The cross-polar cap saturation in GUMICS-4 during high solar wind driving. J. Geophys. Res. Space Phys. 123, 3320–3332. https://doi.org/10.1002/2017JA025054.
Lakhina, G.S., Tsurutani, B.T., 2016. Geomagnetic storms: historical perspective to modern view. Geosci. Lett. 3, 5. doi:10.1186/s40562-016-0037-4.
Landau, L.D., Lifshitz, E.M., 1987. Fluid Mechanics, 2nd ed. Oxford Pergamon Press.
Lanzerotti, L.J., 1992. Comment on "Greatmagnetic storms" by Tsurutani et al. Geophys. Res. Lett. 19, 1991–1992. doi:10.1029/92gl02238.
Lanzerotti, L.J., Medford, L.V., Maclennan, C.G., Kraus, J.S., Kappenman, J., Radasky, W., 2001. Trans-Atlantic geopotentials during the July 2000 solar event and geomagnetic storm. Sol. Phys. 204, 351–359. doi:10.1023/A:1014289410205.
Lanzerotti, L.J., 2017. Space weather: historical and contemporary perspectives. Space Sci. Rev. 212, 1253–1270. https://doi.org/10.1007/s11214-017-0408-y.
Lario, D., Sanahuja, B., Heras, A.M., 1998. Energetic particle events: efficiency of interplanetary shocks as 50 keV < E < 100 MeV proton accelerators. Astrophys. J. 509, 415–434.
Lario, D., Decker, R.B., Roelof, E.C., Vinas, A.-F., 2015. Energetic particle pressure in intense ESP events. J. Phys: Conf. Ser. 642, 012014. doi:10.1088/1742-6596/642/1/012014.

Lastovicka, J., 2021. Long-term trends in the upper atmosphere. In: Wang, W., Zhang, Y., Paxton, L.J. (Eds.), Upper Atmosphere Dynamics and Energetics. https://doi.org/10.1002/9781119815631.ch17.
Latocha, M., Beck, P., Rollet, S., 2009. AVIDOS–a software package for European accredited aviation dosimetry. Radiat. Prot. Dosimetry 136 (4), 286–290. doi:10.1093/rpd/ncp126.
Lavraud, B., Borovsky, J.E., Ridley, A.J., Pogue, E.W., Thomsen, M.F., Reme, H., Fazakerley, A.N., Lucek, E.A., 2007. Strong bulk flow plasma acceleration in Earth's magnetosheath: a magnetic slingshot effect? Geophys. Res. Lett. 34, L14102. doi:10.1029/2007GL030024.
Lavraud, B., Borovsky, J.E., 2008. Altered solar wind-magnetoshpere interaction at low Mach numbers: coronal mass ejections. J. Geophys. Res. 113, A00B08. doi:10.1029/2008JA013192.
Lee, C.O., et al., 2017. MAVEN observations of the solar cycle 24 space weather conditions at Mars. J. Geophys. Res. Space Phys. 122, 2768–2794. doi:10.1002/2016JA023495.
Lee, C.O., et al., 2018. Observations and impacts of the 10 September 2017 solar events at Mars: an overview and synthesis of the initial results. Geophys. Res. Lett. 45, 8871–8885. https://doi.org/10.1029/2018GL079162.
Leske, R.A., Mewaldt, R.A., Stone, E.C., von Rosenvinge, T.T., 2001. Observations of geomagnetic cutoff variations during solar energetic particle events and implications for the radiation environment at the Space Station. J. Geophys. Res. 106 (A12), 30011–30022. doi:10.1029/2000JA000212.
Li, X., Temerin, M., Tsurutani, B.T., Alex, S., 2006. Modeling the 1–2 September 1859 super magnetic storm. Adv. Space Res. 38, 273–279.
Li, X., Roth, I., Temerin, M., Wygant, J.R., Hudson, M.K., Blake, J.B., 1993. Simulation of the prompt energization and transport of radiation belt particles during the March 24, 1991. SSC 20 (22), 2423–2426. https://doi.org/10.1029/93GL02701.
Li, X., Temerin, M., Tsurutani, B.T., Alex, S., 2005. Modeling of 1-2 September 1859 super magnetic storm. Adv. Space Res. 38 (2), 273–279. doi:10.1016/j.asr.2005,06,070.
Liemohn, M.W., et al., 2021. RMSE is not enough: guidelines to robust data-model comparisons for magnetospheric physics. J. Atmos. Sol. Terr. Phys. 218. https://doi.org/10.1016/j.jastp.2021.105624.
Lillis, R.J., Robbins, S., Manga, M., Halekas, J.S., Frey, H.V., 2013. Time history of the Martian dynamo from crater magnetic field analysis. J. Geophys. Res. Planets 118, 1488–1511. doi:10.1002/jgre.20105.
Liou, K., Sotirelis, T., Richardson, I., 2018. Substorm occurrence and intensity associated with three types of solar wind structure. J. Geophys. Res. Space Phys. 123, 485–496. https://doi.org/10.1002/2017JA024451.
Liu, Y., Luhmann, J., Kajdi, P., et al., 2014. Observations of an extreme storm in interplanetary space caused by successive coronal mass ejections. Nat. Commun. 5, 3481. https://doi.org/10.1038/ncomms4481.
Liu, J., Wang, W., Burns, A., Yue, X., Zhang, S., Zhang, Y., Huang, C., 2016. Profiles of ionospheric storm-enhanced density during the 17 March 2015 great storm. J. Geophys. Res. Space Physics 121, 727–744. doi:10.1002/2015JA021832.
Lockwood, M., Hapgood, M., 2007. The rough guide to the Moon and Mars. A&G 48, 6.11–6.17.
Lopez, R.E., 1987. Solar cycle invariance in solar wind proton temperature relationships. J. Geophys. Res. 92 (A10), 11189–11194. doi:10.1029/JA092iA10p11189.
Love, J.J., Rigler, E.J., Pulkkinen, A., Riley, P., 2015. On the lognormality of historical magnetic storm intensity statistics: Implications for extreme-event probabilities. Geophys. Res. Lett. 42, 6544–6553. doi:10.1002/2015GL064842.
Love, J.J., 2020. Some experiments in extreme-value statistical modeling of magnetic superstorm intensities. Space Weather 18, e2019SW002255. https://doi.org/10.1029/2019SW002255.

Lugaz, N., Farrugia, C.J., Huang, C.-L., Spence, H.E., 2015. Extreme geomagnetic disturbances due to shocks within CMEs. Geophys. Res. Lett. 42. doi:10.1002/2015GL064530.
Lugaz, N., Farrugia, C.J., C.-L., H., Winslow, R.M., Spence, H.E., Schwadon, N.A., 2016. Earth's magnetosphere and outer radiation belt under sub-Alfvenic solar wind. Nature Comm 7, 13001. doi:10.1038/ncomms13001.
Luhmann, J.G., Ledvina, S.A., Russell, C.T., 2004. Induced magnetospheres. Adv. Space Res. 33 (11), 1905–1912. https://doi.org/10.1016/j.asr.2003.03.031.
Luhmann, J.G., Ledvina, S.A., Krauss-Varban, D., Odstrcil, D., Riley, P., 2007. A heliospheric simulation-based approach to SEP source and transport modeling. Adv. Space Res. 40 (3), 295–303. https://doi.org/10.1016/j.asr.2007.03.089.
Luhmann, J.G., et al., 2017. Modeling solar energetic particle events using ENLIL heliosphere simulations. Space Weather 15, 934–954. doi:10.1002/2017SW001617.
Luhman, J.G., Mays, M.L., Li, Y., Lee, C.O., Bain, H., Odstrcil, D., Mewaldt, R.A., Cohen, C.M.S., Larson, D., Petrie, G., 2018. Shock connectivity and the late cycle 24 solar energetic particle events in July and September 2017. Space Weather 16, 557–568.
Maehara, H., Shibayama, T., Notsu, S., Notsu, Y., Nagao, T., Kusaba, S., Honda, S., Nogami, D., Shibata, K., 2012. Superflares on solar-type stars. Nature 485, 478–481. doi:10.1038/nature11063.
Maehara, H., Shibayama, T., Notsu, Y., et al., 2015. Statistical properties of superflares on solar-type stars based on 1-min cadence data. Earth Planet Sp 67, 59. https://doi.org/10.1186/s40623-015-0217-z.
Mann, I., Ozeke, L., Murphy, K., et al., 2016. Explaining the dynamics of the ultra-relativistic third Van Allen radiation belt. Nat. Phys. 12, 978–983. https://doi.org/10.1038/nphys3799.
Mannucci, A.J., Tsurutani, B.T., Iijima, B.A., Komjathy, A., Saito, A., Gonzalez, W.D., Guarnieri, F.L., Kozyra, J.U., Skoug, R., 2005. Dayside global ionospheric response to the major interplanetary events of October 29-30, 2003 "Halloween Storms". Geophys. Res. Lett. 32, L12S02. doi:10.1029/2004GL021467.
Marshall, J.E.A., Lakin, J., Troth, I., Wallace-Johnson, S.M., 2020. UV-B radiation was the Devonian-Carboniferous boundary terrestrial extinction kill mechanism. Sci. Adv. 6 (22), eaba0768. doi:10.1126/sciadv.aba0768.
Marsh, M.S., Dalla, S., Dierckxsens, M., Laitinen, T., Crosby, N.B., 2015. SPARX: a modeling system for solar energetic particle radiation space weather forecasting. Space Weather 13, 386–394. doi:10.1002/2014SW001120.
Marsden, P.L., Berry, J.W., Fieldhouse, P., Wilson, J.G., 1956. Variation of cosmic-ray nucleon intensity during the disturbance of 23 February 1956. J. Atmos. Terr. Phys. 8, 278–281.
Matsumoto, T., Shiota, D., Kataoka, R., Miyahara, H., Miyake, S., 2019. A dynamical model of the heliosphere with the Adaptive Mesh Refinement. J. Phys.: Conf. Ser. 1225, 012008. doi:10.1088/1742-6596/1225/1/012008.
Matsumoto, T., Suzuki T.K., 2014. Connecting the Sun and the solar wind: the self-consistent transition of heating mechanisms, Mon. Not. R. Astron. Soc. 440 (2), 971–986. https://doi.org/10.1093/mnras/stu310.
Matthia, D., Berger, T., Mrigakshi, A.I., Reitz, G., 1 February 2013. A ready-to-use galactic cosmic ray model. Adv. Space Res. 51 (3), 329–338.
Matzka, J., Stolle, C., Yamazaki, Y., Bronkalla, O., Morschhauser, A., 2021. The geomagnetic Kp index and derived indices of geomagnetic activity. Space Weather 19, e2020SW002641. https://doi.org/10.1029/2020SW002641.
Mauk, B.H., Fox, N.J., 2010. Electron radiation belts of the solar system. J. Geophys. Res. 115, A12220. doi:10.1029/2010JA015660.
Mauk, B.H., Cohen, I.J., Westlake, J.H., Anderson, B.J., 2016. Modeling magnetospheric energetic particle escape across Earth's magnetopause as observed by the MMS mission. Geophys. Res. Lett. 43, 4081–4088. doi:10.1002/2016GL068856.

Mayaud, P.N., 1980. Derivation, meaning, and use of geomagnetic indices. Geophys. Monogr. Ser. 22, 40–85. doi:10.1029/GM022.

McComas, D.J., Elliott, H.A., Schwadron, N.A., Gosling, J.T., Skoug, R.M., Goldstein, B.E., 2003. The three-dimensional solar wind around solar maximum. Geophys. Res. Lett. 30, 1517. doi:10.1029/2003GL017136 10.

McComas, D.J., Ebert, R.W., Elliott, H.A., Goldstein, B.E., Gosling, J.T., Schwadron, N.A., Skoug, R.M., 2008. Weaker solar wind from the polar coronal holes and the whole Sun. Geophys. Res. Lett. 35, L18103. doi:10.1029/2008GL034896.

McHarg, M.G., Hampton, D.L., Stenbaek-Nielsen, H.C., 1998. Fast photometry of flickering in discrete auroral arcs. Geophys. Res. Lett. 25 (14), 2637–2640. doi:10.1029/98 GL01972.

McIlwain, C.E., 1961. Coordinates for mapping the distribution of magnetically trapped particles. J. Geophys. Res. 66 (11), 3681–3691.

Mekhald, F., et al., 2015. Multiradionuclide evidence for the solar origin of the cosmic-ray events of AD 774/5 and 993/4. Nat. Comm.. doi:10.1038/ncomms9611.

Meredith, N.P., Horne, R.B., Isles, J.D., Rodriguez, J., 2015. Extreme relativistic electron fluxes at geosynchronous orbit: analysis of GOES E >2 MeV electrons. Space Weather 13, 170–184. doi:10.1002/2014SW001143.

Mertens, C.J., Tobiska W.K., Bouwer D., Kress B.T., Wiltberger M., Solomon S.C., Murray J.J., 2009. Development of the Nowcast of Atmospheric Ionizing Radiation for Aviation Safety (NAIRAS) model, AIAA 2009-3633-978.

Mewaldt, R.A., et al., 2005. Proton, helium, and electron spectra during the large solar particle events of October-November 2003. J. Geophys. Res. 110, A09S18.

Millan, R.M., Thorne, R.M., 2007. Review of radiation belt relativistic electron losses. J. Atmos. Sol. Terr. Phys. 69, 362–377.

Minamoto, Y., Fujita, S., Hara, M., 2015. Frequency distributions of magnetic storms and SI+SSC-derived records at Kakioka, Memambetsum and Kanoya. Earth Planets Space 67, 191. doi:10.1186/s40623-015-0362-4.

Miyake, F., Nagaya, K., Masuda, K., Nakamura, T., 2012. A signature of cosmic-ray increase in AD 774-775 from tree rings in Japan. Nature 486, 240–242.

Miyahara, H., Masuda, K., Muraki, Y., et al., 2004. Cyclicity of solar activity during the Maunder Minimum deduced from radiocarbon content. Sol. Phys. 224, 317–322. https://doi.org/10.1007/s11207-005-6501-5.

Mironova, I.A., Aplin, K.L., Arnold, F., et al., 2015. Energetic particle influence on the Earth's atmosphere. Space Sci. Rev. 194, 1–96. https://doi.org/10.1007/s11214-015-0185-4.

Miyahara, H., Wen, G., Cahalan, R.F., Ohmura, A., 2008. Deriving historical total solar irradiance from lunar borehole temperatures. Geophys. Res. Lett. 35, L02716. doi:10.1 029/2007GL032171.

Miyahara, H., Yokoyama, Y., Yamaguchi, Y., 2009. Influence of the Schwabe/Hale solar cycles on climate change during the Maunder Minimum. Proc. Int. Astron. Union 5 (S264), 427–433. doi:10.1017/S1743921309993048.

Miyahara, H., Tokanai, F., Moriya, T., et al., 2021. Gradual onset of the Maunder Minimum revealed by high-precision carbon-14 analyses. Sci. Rep. 11, 5482. https://doi.org/10.1038/s41598-021-84830-5.

Miyake, S., Kataoka, R., Sato, T., 2017. Cosmic ray modulation and radiation dose of aircrews during the solar cycle 24/25. Space Weather 15 (4), 589–605. doi:10.1002/201 6SW001588.

Miyaoka, H., Hirasawa, T., Yumoto, K., Tanaka, Y., 1990. 12. Low latitude aurorae on October 21, 1989, I, Proc. Japan Acad., 66, Ser B., 47–51.

Miyoshi, Y., Morioka, A., Obara, T., Misawa, H., Nagai, T., Kasahara, Y., 2003. Rebuilding process of the outer radiation belt during the November 3: 1993, magnetic storm-NOAA and EXOS-D observations. J. Geophys. Res 108 (A1), 1004. https://doi.org/10.1029/2001ja007542.

Miyoshi, T., Kusano, K., 2005. A multi-state HLL approximate Riemann solver for ideal magnetohydrodynamics. J. Comput. Phys. 208 (1), 315–344. https://doi.org/10.1016/j.jcp.2005.02.017.
Miyoshi, Y., Kataoka, R., 2005. Ring current ions and radiation belt electrons during geomagnetic storms driven by coronal mass ejections and corotating interaction regions. Geophys. Res. Lett. 32, L21105. doi:10.1029/2005GL024590.
Miyoshi, Y., Kataoka, R., 2008. Probabilistic space weather forecast of the relativistic electron flux enhancement at geosynchronous orbit. J. Atmos. Sol. Terr. Phys. 70, 475–481. doi:10.1016/j.jastp.2007.08.066.
Miyoshi, Y., Sakaguchi, K., Shiokawa, K., Evans, D., Albert, J., Connors, M., 2008. Precipitation of radiation belt electrons by EMIC waves, observed from ground and space. Geophys. Res. Lett. 35, L23101. doi:10.1029/2008GL035727.
Miyoshi, Y., Kataoka, R., 2011. Solar cycle variations of outer radiation belt and its relationship to solar wind structure dependences. J. Atm. Solar-Terr. Phys. 73, 77–87.
Miyoshi, Y., Kataoka, R., Kasahara, Y., Kumamoto, A., Nagai, T., Thomsen, M., 2013. High-speed solar wind with southward interplanetary magnetic field causes relativistic electron flux enhancement of the outer radiation belt via enhanced condition of whistler waves. Geophys. Res. Lett. 40, 4520–4525. doi:10.1002/grl.50916.
Miyoshi, Y., Shinohara, I., Takashima, T., Asamura, K., Higashio, N., Mitani, T., Kasahara, S., Yokota, S., Kazama, Y., Wang, S.-. Y., Tam, S.W., Ho, P.T.P, Kasahara, Y., Kasaba, Y., Yagitani, S., Matsuoka, A., Kojima, H., Katoh, H., Shiokawa, K., Seki, K., 2018. Geospace exploration project ERG. Earth Planets Space 70, 101. doi:10.1186/s40623-018-0862-0.
Miyoshi, et al., 2021. Penetration of MeV electrons into the mesosphere accompanying pulsating aurorae. Sci. Rep. 11 (2021) Article number: 13724.
Mozer, F.S., Artemyev, A., Agapitov, O.V., Mourenas, D., Vasko, I., 2016. Near-relativistic electron acceleration by Landau trapping in time domain structures. Geophys. Res. Lett. 43, 508–514. doi:10.1002/2015GL067316.
Mullen, E.G., Gussenhoven, M.S., Ray, K., Violet, M., 1991. A double-peaked inner radiation belt: cause and effect as seen on CRRES. IEEE Trans. Nucl. Sci. 38 (6), 1713–1718. doi:10.1109/23.124167.
Muscheler, R., Adolphi, F., Herbst, K., et al., 2016. The Revised Sunspot Record in Comparison to Cosmogenic Radionuclide-Based Solar Activity Reconstructions. Sol. Phys. 291, 3025–3043. https://doi.org/10.1007/s11207-016-0969-z.
Nakamizo, A., Tanaka, T., Kubo, Y., Kamei, S., Shimazu, H., Shinagawa, H., 2009. Development of the 3-D MHD model of the solar corona-solar wind combining system. J. Geophys. Res. 114, A07109. doi:10.1029/2008JA013844.
Naito, M., Hasebe, N., Shikishima, M., Amano, Y., Haruyama, J., Matis-Lopes, J.A., Kim, K.J., Kodaira, S., 2020. Radiation dose and its protection in the Moon from galactic cosmic rays and solar energetic particles: at the lunar surface and a lava tube. J. Raiol. Prot. 40, 946–961.
Nakagawa, H., 2019. Atmosphere of Mars, In: Yamagishi, A., Kakegawa, T., Usui, T. (Eds.), Astrobiology: From the Origins of Life to the Search for Extraterrestrial Intelligence, Springer, Singapore, pp. 353–366.
Nakamura, R., Baker, D.N., Blake, J.B., Kanekal, S., Klecker, B., Hovestadt, D., 1995. Relativistic electron precipitation enhancements near the outer edge of the radiation belt. Geophys. Res. Lett. 22 (9), 1129–1132.
Nakamura, M., Yoneda, A., Oda, M., Tsubouchi, K., 2015. Statistical analysis of extreme auroral electrojet indices. Earth Planet Space 67, 153. doi:10.1086/s40623-015-0321-0.
Nakamura, S., Ebihara, Y., Fujita, S., Goto, T., Yamada, N., Watari, S., Omura, Y., 2018. Time domain simulation of geomagnetically induced current (GIC) flowing in 500-kV power grid in Japan including a three-dimensional ground inhomogeneity. Space Weather 16, 1946–1959. https://doi.org/10.1029/2018SW002004.

Nakamura, Y., Terada, N., Leblanc, F., Rahmati, A., Nakagawa, H., Sakai, S., et al., 2022. Modeling of diffuse auroral emission at Mars: Contribution of MeV protons. J. Geophys. Res. Space Phys. 127. e2021JA029914. https://doi.org/10.1029/2021JA029914.
Nansen, F., 1911. Northern Mists: Arctic Exploration in Early Times II. Frederick A. Stokes, New York.
NASA, 2014. NASA Space Flight human-system standard volume 1, revision A: Crew health. (NASA-STD-3001, VOLUME 1). Retrieved from https://standards.nasa.gov/standard/nasa/nasa-std-3001-vol-1.
National Research Council, 2008. Severe Space Weather Events: Understanding Societal and Economic Impacts: A Workshop Report. The National Academies Press, Washington, DC. https://doi.org/10.17226/12507.
Newman, M.E.J., 2005. Power laws, Pareto distributions and Zipf's law. Contemp. Phys. 46, 323–351.
Ngwira, C.M., Pulkkinen, A., Mays, M.L., Kuznetsova, M.M., Galvin, A.B., Simunac, K., Baker, D.N., Li, X., Zheng, Y., Glocer, A., 2013a. Simulation of the 23 July 2012 extreme space weather event: what if this extremely rare CME was Earth directed? Space Weather 11, 1–9. doi:10.1002/2013SW000990.
Ngwira, C.M., Pulkkinen, A., Wilder, F.D., Crowley, G., 2013b. Extended study of extreme geoelectric field event scenarios for geomagnetically induced current applications. Space Weather 11, 121–131. doi:10.1002/swe.20021.
Ngwira, C.M., Pulkkinen, A., Kuznetsova, M.M., Glocer, A., 2014. Modeling extreme "Carrington-type" space weather events using three-dimensional global MHD simulations. J. Geophys. Res. Space Phys. 119, 4456–4474. doi:10.1002/2013JA019661.
Nilsson, H., 2011. Heavy ion energization, transport, and loss in the Earth's magnetosphere. In: Liu, W., Fujimoto, M. (Eds.), The Dynamic Magnetosphere. Springer, Dordrecht Heidelberg, London, New York.
Nishimura, Y., Lessard, M.R., Kato, Y., Miyoshi, Y., Grono, E., Partamies, N., Sivadas, N., Hosokawa, K., Fukizawa, M., Samara, M., Michell, R.G., Kataoka, R., Sakanoi, T., Whiter, D.K., 2020. Diffuse and pulsating aurora. Space Sci. Rev. 216 (2020), 4. https://doi.org/10.1007/s11214-019-0629-3.
Nishitani, N., Papitashvili, V.O., Ogawa, T., Sato, N., Yamagishi, H., Yukimatu, A.S., Rich, F.J., 2003. Interhemispheric asymmetry of the high-latitude ionospheric convection on 11-12 May 1999. J. Geophys. Res. 108 (A5), 1184. doi:10.1029/2002JA009680.
Nishizuka, N., Sugiura, K., Kubo, Y., Den, M., Ishii, M., 2018. Deep Flare Net (DeFN) model for solar flare prediction. Astrophys. J. 858, 113.
Nitta, N.V., Mulligan, T., 2017. Earth-affecting coronal mass ejections without obvious low coronal signatures. Sol. Phys. 292, 125. doi:10.1007/s11207-017-1147-7.
NOAA, 2014. Intense Space Weather Storms October 19-November 07, 2003. NOAA National Weather Service, Silver Spring, Md. April 2004, p1.
Nose, M., Taguchi, S., Hosokawa, K., Christon, S.P., McEntire, R.W., Moore, T.E., Collier, M.R., 2005. Overwhelming O+ contribution to the plasma sheet energy density during the October 2003 superstorm: Geotail/EPIC and IMAGE/LENA observations. J. Geophys. Res. 110, A09S24. doi:10.1029/2004JA010930.
Notsu, Y., Maehara H., Honda S., Hawley S.L., Davenport J.R.A., Namekata K., Notsu S., Ikuta K., Nogami D., Shibata K., 2019. Do Kepler superflare stars really include slowly rotating Sun-like stars?-Results using APO 3.5 m Telescope Spectroscopic Observations and Gaia-DR2 Data, 876, 1, 58.
O'Brien, T.P., McPherron, R.L., 2000. An empirical phase space analysis of ring current dynamics: Solar wind control of injection and decay. J. Geophys. Res. 105 (A4), 7707–7719. doi:10.1029/1998JA000437.
O'Brien, T.P., Mazur, J.E., Looper, M.D., 2018. Solar energetic proton access to the magnetosphere during the 10-14 September 2017 particle event. Space Weather 16, 2022–2037. https://10.1029/2018SW001960.

Odstrcil, D., Pizzo, V.J., 1999. Three-dimensional propagation of coronal mass ejections (CMEs) in a structured solar wind flow: 1. CME launched within the streamer belt. J. Geophys. Res. 104 (A1), 483–492. doi:10.1029/1998JA900019.

Ogliore, R.C., Mewaldt, R.A., Leske, R.A., Stone, E.G., von Rosenvinge, T.T., 2001. A direct measurement of the geomagnetic cutoffs for cosmic rays at space station latitudes, Proc. 27th Intl. Cosmic Ray Conf., 10, 4112–4115.

Opher, M., Loeb, A., Drake, J., Toth, G., 2020. A small and round heliosphere suggested by magnetohydrodynamic modelling of pick-up ions. Nat. Astron. 4, 675–683, https://doi.org/10.1038/s41550-020-1036-0.

Parker, E.N., 1957. Sweet's mechanism for merging magnetic fields in conducting fluids. J. Geophys. Res. 62 (4), 509–520. doi:10.1029/JZ062i004p00509.

Parker, E.N., 1958. Dynamics of the interplanetary gas and magnetic fields. Astrophys. J. 128, 664, doi: 10.1086/146579.

Parker, E.N., 1965. The passage of energetic charged particles through interplanetary space. Planet. Space Sci. 13, 9–49.

Pallamraju, D., Chakrabarti, S., 2005. First ground-based measurements of OI 6300 A daytime aurora over Boston in response to the 30 October 2003 geomagnetic storm. Geophys. Res. Lett. 32, L03S10. doi:10.1029/2004GL021417.

Parks, G., Brittnacher, M., Chua, D., Fillingim, M., Germany, G., Spann, J., 2000. Behavior of the aurora during 10-12 May, 1999 when the solar wind nearly disappeared. Geophys. Res. Lett. 27 (24), 4033–4036.

Particle Data Group, 2020. Review of particle physics. Progr. Theor. Exp. Phys. 2020 (8), 083C01. August 2020. https://doi.org/10.1093/ptep/ptaa104.

Paulikas, G.A., Blake, J.B., 1979. Effects of the solar wind on magnetospheric dynamics: Energetic electrons at the synchronous orbit. in Quantitative Modeling of Magnetospheric Processes. Geophys. Monogr. Ser. 21, 180–202 AGU, Washington, D. C.

Petschek, H.E., 1964, Magnetic Field Annihilation. In: Hess, W.N. (Ed.), The Physics of Solar Flares, Proceedings of the AAS-NASA Symposium held 28-30 October, 1963 at the Goddard Space Flight Center, Greenbelt, MD. National Aeronautics and Space Administration, Science and Technical Information Division, Washington, DC, pp. 425.

Pizzo, V., 1978. A three-dimensional model of corotating streams in the solar wind 1. Theoretical Foundations. J. Geophys. Res. 83 (A12), 5563–5572.

Pollack, J.B., Kasting, J.F., Richardson, S.M., Poliakoff, K., 1987. The case for a wet, warm climate on early Mars. Icarus 71, 203–224.

Poluianov, S.V., Kovaltsov, G.A., Usoskin, I.G., 2019. Extreme Solar Particle Storms ed F. Miyake, I. G. Usoskin and S. Poluianov. IOP Publishing, Bristol.

Pomoell, J., Poedts, S., 2018. EUHFORIA: European heliospheric forecasting information asset. J. Space Weather Space Clim. 8, A35. https://doi.org/10.1051/swsc/2018020.

Pontin, D.I., Priest, E.R., Galsgaard, K., 2013. On the nature of reconnection at a solar coronal null point above a Separatrix Dome. Astrophys. J. 774, 154.

Pulkkinen, A., Lindahl, S., Viljanen, A., Pirjola, R., 2005. Geomagnetic storm of 29-31 October 2003: geomagnetically induced currents and their relation to problems in the Swedish high-voltage power transmission system. Space Weather 3, S08C03. doi:10.1029/2004SW000123.

Pulkkinen, A., Kataoka, R., Watari, S., Ichiki, M., 2010. Modeling geomagnetically induced currents in Hokkaido, Japan. Adv. Space Res. 46 (9), 1087–1093. doi:10.1016/j.asr.2010.05.024.

Pulkkinen, A., Bernabeu, E., Eichner, J., Beggan, C., Thomson, A.W.P., 2012. Generation of 100-year geomagnetically induced current scenarios. Space Weather 10, S04003. doi:10.1029/2011SW000750.

Pulkkinen, A., 2017. Introduction to NASA Living With a Star (LWS) Institute GIC Working Group Special Collection. Space Weather 15, 738–740. doi:10.1002/2016SW001537.

Pulkkinen, A., Bernabeu, E., Thomson, A., Viljanen, A., Pirjola, R., Boteler, D., Eichner, J., Cilliers, P.J., Welling, D., Savani, N.P., Weigel, R.S., Love, J.J., Balch, C., Ngwira, C.M., Crowley, G., Schultz, A., Kataoka, R., Anderson, B., Fugate, D., Simpson, J.J., MacAlester, M., 2017. Geomagnetically induced currents: science, engineering and applications readiness. Space Weather 15, 828–856. doi:10.1002/2016SW001501.

Rajesh, P.K., Lin, C.H., Lin, C.Y., Chen, C.H., Liu, J.Y., Matsuo, T., et al., 2021. Extreme positive ionosphere storm triggered by a minor magnetic storm in deep solar minimum revealed by FORMOSAT-7/COSMIC-2 and GNSS observations. J. Geophys. Res. Space Phys. 126, e2020JA028261. https://doi.org/10.1029/2020JA028261.

Ramirez, R.M., Kopparapu, R., Zugger, M.E., Robinson, T.D., Freedman, R., Kasting, J.F., 2014. Warming early Mars with CO_2 and H_2. Nat. Geosci. 7, 59–63. doi:10.1038/NGEO2000.

Reames, D.V., 1999. Solar energetic particles: is there time to hide? Radiat. Meas. 30, 297–308.

Reames, D.V., 2009. Solar energetic-particle release times in historic ground-level events. Astrophys. J. 706 (1), 844 Number.

Reames, D.V., 2010. Remote Sensing of Magnetic-Cloud Topology. Sol Phys 265, 187–195. https://doi.org/10.1007/s11207-010-9527-2.

Reames, D.V., 2013. The two sources of solar energetic particles. Space Sci. Rev. 175, 53–92. doi:10.1007/s11214-013-9958-9.

Reames, D., Ng, C.K., Tylka, A.J., 2013. Spatial distribution of solar energetic particles in the inner heliosphere. Solar Phys 285, 233–250. doi:10.1007/s11207-012-0038-1.

Reames, D.V., 2017. Solar energetic particles, a modern primer on understanding sources, acceleration and propagationLecture Notes in Physics 932. Springer, Heidelberg.

Rees, M.H., 1984. Excitation of O(1S) and emission of 5577 A radiation in aurora. Planet. Space Sci. 32 (3), 373–378.

Rees, M.H., 1989. Physics and chemistry of the upper atmosphere. Cambridge Atmospheric and Science Series, Cambridge University Press.

Reeves, G.D., et al., 2013. Electron acceleration in the heart of the Van Allen Radiation Belts. Science 341 (6149), 991–994. doi:10.1126/science.1237743.

Reiss, M.A., MacNeice, P.J., Mays, L.M., Arge, C.N., Mostl, C., Nikolic, L., Amerstorfer, T., 2019. Forecasting the ambient solar wind with numerical models. I. On the implementation of an operational framework. ApJS 240, 35.

Reiss, M.A., MacNeice, P.J., Muglach, K., Arge, C.N., Mostl, C., Riley, P., Hinterreiter, J., Bailey, R.L., Weiss, A.J., Owens, M.J., Amerstorfer, T., Amerstorfer, U., 2020. Forecasting the ambient solar wind with numerical models. II. An adaptive prediction system for specifying solar wind speed near the Sun. Astrophys. J. 891, 165.

Rhodes, D.J., Farrell, W.M., McLain, J.L., 2020. Tribocharging and electrical grounding of a drill in shadowed regions of the Moon. Adv. Space Res. 66, 753–759. doi:10.1016/j.asr.2020.05.005.

Ribas, I., Guinan, E.F., Güdel, M., Audard, M., 2005. Evolution of the Solar Activity over Time and Effects on Planetary Atmospheres. I. High-Energy Irradiances (1-1700 Å). ApJ 622, 680.

Richardson, I.G., Cane, H.V., 1995. Regions of abnormally low proton temperature in the solar wind (1965-1991) and their association with ejecta. J. Geophys. Res. 100 (A12), 23397–23412. doi:10.1029/95JA02684.

Richardson, I.G., Webb, D.W., Zhang, J., Berdichevsky, D., Biesecker, D.A., Kasper, J.C., Kataoka, R., Steinberg, J., Thompson, B.J., Wu, C.-C., Zhukov, A., 2006. Major geomagnetic storms (Dst < -100 nT) generated by corotating interaction regions. J. Geophys. Res. 111, A07S09. doi:10.1029/2005JA011476.

Richardson, I.G., et al., 2014. 25 MeV proton events observed by the high energy telescopes on the STEREO A and B Spacecraft and/or at earth during the first seven years of the STEREO mission. Solar Phys. 289, 3059–3107.

Richardson, I.G., Mays, M.L., Thompson, B.J., 2018. Prediction of solar energetic particle event peak proton intensity using a simple algorithm based on CME speed and direction and observations of associated solar phenomena. Space Weather 16, 1862–1881. https://doi.org/10.1029/2018SW002032.
Richmond, A.D., 1995. Ionospheric electrodynamics using magnetic apex coordinates. J. Geomag. Geoelectr. 47, 191–212.
Richmond, N.C., Hood, L.L., 2008. A preliminary global map of the vector lunar crustal magnetic field based on Lunar Prospector magnetometer data. J. Geophys. Res. 113, E02010. doi:10.1029/2007JE002933.
Ridley, A.J., 2007. Alfven wings at Earth's magnetopause under strong interplanetary magnetic fields. Ann. Geophys. 25 (2), 533–542.
Riley, P., Linker, J.A., Mikic, Z., 2002. Modeling the heliospheric current sheet: solar-cycle variations. J. Geophys. Res. 107 (A7), 1136. doi:10.1029/2001JA000299.
Riley, P., 2012. On the probability of occurrence of extreme space weather events. Space Weather 10, S02012. doi:10.1029/2011SW000734.
Riley, P., Love, J.J., 2017. Extreme geomagnetic storms: probabilistic forecasts and their uncertainties. Space Weather 15, 53–64. https://doi.org/10.1002/2016SW001470.
Riley, P., Baker, D., Liu, Y.D., Verronen, P., Singer, H., Gudel, M., 2018. Extreme space weather events: from cradle to grave. Space Sci. Rev. 214, 21. https://doi.org/10.1007/s11214-017-0456-3.
Rishbeth, H., Shea, M.A., Smart, D.F., 2009. The solar-terrestrial event of 23 February 1956. Adv. Space Res. 44 (10), 1096–1106.
Rohen, G., et al., 2005. Ozone depletion during the solar proton events of October/November 2003 as seen by SCIAMACHY. J. Geophys. Res. 110, A09S39. doi:10.1029/2004JA010984.
Robbrecht, E., Patsourakos, S., Vourlidas, A., 2009. No trace left behind: Stereo observation of a coronal mass ejection without low coronal signatures. Astrophys. J. 701 (1), 283.
Rufenach, C.L., Martin, R.F., Sauer, H.H., 1989. A study of geosynchronous magnetopause crossings. J. Geophys. Res. 94 (A11), 15125–15134. doi:10.1029/JA094iA11p15125.
Ruffolo, D., 1991. Interplanetary transport of decay protons from solar flare neutrons. Astrophys. J. 382, 688–698.
Russell, C.T., 1980. Planetary magnetism. Rev. Geophys. 18 (1), 77–106. doi:10.1029/RG018i001p00077.
Russell, C.T., Luhmann, J.G., 1997. Mercury: Magnetic field and magnetosphereEncyclopedia of Planetary Science. Encyclopedia of Earth Science. Springer, Dordrecht. https://doi.org/10.1007/1-4020-4520-4_253.
Russell, C.T., 2004. Outer planet magnetospheres: a tutorial. Adv. Space Res. 33 (11), 2004–2020.
Russell., C.T., et al., 2013. The very unusual interplanetary corona mass ejections of 2012 July 23: a blast wave mediated by solar energetic particles. Astrophys. J. 770 (38), 5.
Russell, C.T., McPherron, R.L., 1973. Semiannual variation of geomagnetic activity. J. Geophys. Res. 78, 92.
Sagan, C., Pollack, J., 1969. Windblown dust on Mars. Nature 223, 791–794. https://doi.org/10.1038/223791a0.
Sagan, C., Mullen, G., 1972. Earth and Mars: evolution of the atmospheres and surface temperatures. Science 177 (4043), 52–56. doi:10.1126/science.177.4043.52.
Saito, T., 1969. Geomagnetic pulsations. Space Sci. Rev. 10, 319–412.
Saito, Y., Nishino, M.N., Fujimoto, M., Yamamoto, T., Yokota, S, Tsunakawa, H, et al., 2012. Simultaneous observation of the electron acceleration and ion deceleration over lunar magnetic anomalies. Earth Planets Space 64, 83–92.
Sakaguchi, K., Miyoshi, Y., Saito, S., Nagatsuma, T., Seki, K., Murata, K.T., 2013. Relativistic electron flux forecast at geostationary orbit using Kalman filter based on multivariate autoregressive model. Space Weather 11, 79–89, doi:10.1002/swe.20020.

Sakaguchi, K., Nagatsuma, T., Reeves, G.D., Spence, H.E., 2015. Prediction of MeV electron fluxes throughout the outer radiation belt using multivariate autoregressive models. Space Weather 13, 853–867, doi:10.1002/2015SW001254.

Sakai, S., Seki, K., Terada, N., Shinagawa, H., Sakata, R., Tanaka, T., Ebihara, Y., 2021. Effects of the IMF direction on atmospheric escape from a Mars-like planet under weak intrinsic magnetic field conditions. J. Geophys. Res. Space Phys. 126. e2020JA028485. https://doi.org/10.1029/2020JA028485.

Sakata, R., Seki, K., Sakai, S., Terada, N., Shinagawa, H., Tanaka, T., 2020. Effects of an intrinsic magnetic field on ion loss from ancient Mars based on multispecies MHD simulations. J. Geophys. Res. Space Phys. 125. e2019JA026945. https://doi.org/10.1029/2019JA026945.

Sato, T., Kataoka, R., Shiota, D., Kubo, Y., Ishii, M., Yasuda, H., Miyake, S., Park, I., Miyoshi, Y., 2018b. Real-time and automatic analysis program for WASAVIES: warning system of aviation exposure to solar energetic particles. Space Weather 16. https://doi.org/10.1029/2018SW001873.

Sato, T., Iwamoto, Y., Hashimoto, S., Ogawa, T., Furuta, T., Abe, S., Kai, T., Tsai, P.E., Matsuda, N., Iwase, H., Shigyo, N., Sihver, L., Niita, K., 2018c. Features of particle and heavy ion transport code system PHITS Version 3.02. J. Nucl. Sci. Technol. 55, 684–690.

Sandford, B.P., 1961. Enhancement of night-sky molecular nitrogen emission in polar cap absorption events. Nature 4772, 245–246.

Sandford, B.P., 1963. Optical studies of particle bombardment in polar cap absorption events. Planet Space Sci. 10, 195–213.

Schneider, N.M., et al., 2015. Discovery of diffuse aurora on Mars. Science 350 (6261), aad0313. doi:10.1126/science.aad0313.

Schaefer, B.E., King, J.R., Deliyannis, C.P., 2000. Superflares on Ordinary Solar-Type Stars. Astrophys. J. 529 (2), 1026.

Schneider, N.M., Jain, S.K., Deighan, J., Nasr, C.R., Brain, D.A., Larson, D., et al., 2018. Global aurora on Mars during the September 2017 space weather event. Geophys. Res. Lett. 45, 7391–7398. https://doi.org/10.1029/2018GL077772.

Schrijver, C.J., 2007. A characteristic magnetic field pattern associated with all major solar flares and its use in flare forecasting. Astrophys. J. 655, L117–L120.

Schultz, M., Lanzerotti, L.J., 1974. Particle diffusion in the radiation belts, Physics and Chemistry in Space 7. Springer-Verlag, Berlin Heidelberg.

Schwadron, N.A., et al., 2010. Earth-Moon-Mars radiation environment module framework. Space Weather 8, S00E02. doi:10.1029/2009SW000523.

Schwadron, N.A., Rahmanifard, F., Wilson, J., Jordan, A.P., Spence, H.E., Joyce, C.J., et al., 2018. Update on the worsening particle radiation environment observed by CRaTER and implications for future human deep-space exploration. Space Weather 16, 289–303. https://doi.org/10.1002/2017SW001803.

Sckopke, N., 1966. A general relation between the energy of trapped particles and the disturbance field near the Earth. J. Geophys. Res. 71 (13), 3125–3130. doi:10.1029/JZ071i013p03125.

Sedov, L.I., 1946. The movement of air in a strong explosion. Dokl Akad Nauk SSSR 52, 17.

Sergeyeva, N., Gvishiani, A., Soloviev, A., Zabarinskaya, L., Krylova, T., Nisilevich, M., Krasnoperov, R., 2021. Historical K index data collection of Soviet magnetic observatories, 1957-1992. Earth Syst. Sci. Data 13, 1987–1999. https://doi.org/10.5194/essd-13-1987-2021.

Shea, M.A., Smart, D.F., Allen, J.H., Wilkinson, D.C., 1992. Spacecraft problems in association with episodes of intense solar activity and related terrestrial phenomena during March 1991. IEEE Trans. Nucl. Sci. 39 (6), 1754–1760.

Shea, M.A., Smart, D.F., 2006. Compendium of the eight articles on the Carrington Event attributed to or written by Elias Loomis in the American Journal of Science, 1859-1861. Adv. Space Res. 38 (2), 313–385 2006.

Shepherd, S.G., 2007. Polar cap potential saturation: observations, theory, and modeling. J. Atmos. Sol. Terr. Phys. 69 (3), 234–248. doi:10.1016/j.jastp.2006.07.022.
Shue, J.-H., et al., 1998. Magnetopause location under extreme solar wind conditions. J. Geophys. Res. 103 (17), 691.
Shibata, K., Isobe, H., Hillier, A., Choudhuri, A.R., Maehara, H., Ishii, T.T., Shibayama, T., Notsu, S., Notsu, Y., Nagao, T., Honda, S., Nogami, D., 2013. Can superflares occur on our Sun? Publ. Astron. Soc. Japan 65, 49.
Shiokawa, K., Ogawa, T., Oya, H., Rich, F.J., Yumoto, K., 2001. A stable auroral red arc observed over Japan after an interval of very weak solar wind. J. Geophys. Res. 106 (A11), 26091–26101.
Shiokawa, K., Ogawa, T., Kamide, Y., 2005. Low-latitude auroras observed in Japan: 1999–2004. J. Geophys. Res. 110, A05202. doi:10.1029/2004JA010706.
Shiokawa, K., Miyoshi, Y., Brandt, P.C., Evans, D.S., Frey, H.U., Goldstein, J., Yumoto, K., 2013. Ground and satellite observations of low-latitude red auroras at the initial phase of magnetic storms. J. Geophys. Res. Space Phys. 118, 256–270. doi:10.1029/2012JA018001.
Shiota, D., Isobe, H., Chen, P.F., Yamamoto, T.T., Sakajiri, T., Shibata, K., 2005. Self-consistent magnetohydrodynamic modeling of a coronal mass ejection, coronal dimming, and a giant cusp-shaped arcade formation. Astrophys. J. 634 (2005), 663–678. https://doi.org/10.1086/496943.
Shiota, D., Kataoka, R., 2016. Magnetohydrodynamic simulation of interplanetary propagation of multiple coronal mass ejections with internal magnetic flux rope (SUSANOO-CME). Space Weather 14, 56–75. doi:10.1002/2015SW001308.
Shoji, M., Miyoshi, Y., Kistler, L.M., Asamura, K., Matsuoka, A., Kasaba, Y., Matsuda, S., Kasahara, Y., Shinohara, I., 2021. Discovery of proton hill in the phase space during interactions between ions and electromagnetic ion cyclotron waves. Sci. Rep. 11, 13480. doi:10.1038/s41598-021-92641-0.
Shurshakov, V.A., et al., 1999. Solar particle events observed on MIR station. Radiat. Meas. 30 (3), 317–325. https://10.1016/S1350-4487(99)00058-X.
Sigernes, F., Holmes, J.M., Dyrland, M., Lorentzen, D.A., Svenoe, T., Heia, K., Aso, T., Chernouss, S., Deehr, C.S., 2008. Sensitivity calibration of digital colour cameras. Opt. Express 16 (20), 15623–15632.
Silverman, S.M., 2008. Low-latitude auroras: the great aurora of 4 February, 1872. J. Atmos. Sol. Terr. Phys. 70, 1301–1308.
Silverman, S.M., Cliver, E.W., 2001. Low-latitude auroras: the magnetic storm of 14-15 May 1921. J. Atmos. Sol. Terr. Phys. 63, 523–535.
Simpson, F., Bahr, K., 2005. Practical Magnetotellurics. Cambridge University Press, Cambridge. doi:10.1017/CBO9780511614095.
Siscoe, G.L., Erickson, G.M., Sonnerup, B.U.O., Maynard, N.C., Schoendorf, J.A., Siebert, K.D., Weimer, D.R., White, W.W., Wilson, G.R., 2002. Hill model of transpolar potential saturation: comparisons with MHD simulations. J. Geophys. Res. 107 (A6), 1075. doi: 10.1029/2001JA000109.
Siscoe, G., Raeder, J., Ridley, A.J., 2004. Transpolar potential saturation models compared. J. Geophys. Res. 109, A09203. doi:10.1029/2003JA010318.
Siscoe, G., Crooker, N.U., Clauer, C.R., 2006. Dst of the Carrington storm of 1859. Adv. Space Res. 38, 173–179.
Shoda, M., Yokoyama, T., Suzuki, T.K., 2018. A self-consistent model of the coronal heating and solar wind acceleration including compressible and incompressible heating processes. Astrophys. J. 2 (190), 853. doi:10.3847/1538-4357/aaa3e1.
Smart, D.F., Shea, M.A., 2005. A review of geomagnetic cutoff rigidities for earth-orbiting spacecraft. Adv. Space Res. 36, 2012–2020.
Smart, D.F., Shea, M.A., 2009. Fifty years of progress in geomagnetic cutoff rigidity determinations. Adv. Space Res. 44 (10), 1107–1123, https://doi.org/10.1016/j.asr.2009.07.005.

Stone, E.C., Cummings, A.C., McDonald, F.B., Heikkila, B.C., Lal, N., Webber, W.R., 2013. Voyager 1 observes low-energy galactic cosmic rays in a region depleted of heliospheric ions. Science 341 (6142), 150–153. https://doi.org/10.1126/science.1236408.

Stormer, C., 1955. The Polar Aurora. Claredon Press, Oxford.

Strickland, D.J., Daniell, R.E., Jasperse, J.R., Basu, B., 1993. Transport-theoretic model for the electron-proton-hydrogen atom aurora: 2. Model results. J. Geophys. Res. 98 (A12), 21533–21548. doi:10.1029/93JA01645.

Strong, A.W., Moskalenko, I.V., Ptuskin, V.S., 2007. Cosmic-ray propagation and interactions in the galaxy, 57, 285–327, https://doi.org/10.1146/annurev.nucl.57.090506.123011.

Subbotin, D., Shprits, Y., Ni, B., 2010. Three-dimensional VERB radiation belt simulations including mixed diffusion. J. Geophys. Res. 115, A03205, doi:10.1029/2009JA015070.

Suganuma Y. et al., 2021. Formal ratification of the Global Boundary Stratotype Section and Point (GSSP) for the Chibanian Stage and Middle Pleistocene Subseries of the Quaternary System: the Chiba Section, Japan, Episodes, https://doi.org/10.18814/epiiugs/2020/020080.

Sugiura, M., 1964. Hourly values of equatorial Dst for the IGY. Ann. Int. Geophys. Year 35, 9 Pergamon Press, Oxford.

Sugiura, M., Kamei, T., 1991. Equatorial Dst index 1957-1986. IAGA Bull 40 IUGG, Paris.

Sukhodolov, T., Usoskin, I., Rozanov, E., et al., 2017. Atmospheric impacts of the strongest known solar particle storm of 775 AD. Sci Rep 7, 45257. https://doi.org/10.1038/srep45257.

Summers, D., Thorne, R.M., Xiao, F., 1998. Relativistic theory of wave-particle resonant diffusion with application to electron acceleration in the magnetosphere. J. Geophys. Res. 103 (A9), 20487–20500. doi:10.1029/98JA01740.

Suzuki, T.K., Inutsuka, S., 2006. Solar winds driven by nonlinear low-frequency Alfven waves from the photosphere: parametric study for fast/slow winds and disappearance of solar winds. J. Geophys. Res. 111, A06101. doi:10.1029/2005JA011502.

Svalgaard, L., Schatten, K.H., 2016. Reconstruction of the sunspot group number: The backbone method. Solar Phys. 291 (2016), 2653–2684. doi:10.1007/s11207-015-0815-8.

Sweet, P.A., 1958. The neutral point theory of solar flares. Electromagnetic Phenomena in Cosmical Physics 6, Cambridge University Press, 123.

Szabo, A., 1994. An improved solution to the "Rankine-Hugoniot" problem. J. Geophys. Res. 99 (A8), 14737–14746. doi:10.1029/94JA00782.

Takahashi, T., Mizuno, Y., Shibata, K., 2016. Scaling relations in coronal mass ejections and energetic proton events associated with solar superflares. Astrophys. J. Lett. 833 (1), L8 Number.

Takasao, S., Shibata, K., 2016. Above-the-loop-top oscillation and quasi-periodic coronal wave generation in solar flares. Astrophys. J. 823, 150.

Tanaka, T., 2007. Magnetosphere–Ionosphere Convection as a Compound System. Space Sci. Rev. **133**, 1–72. https://doi.org/10.1007/s11214-007-9168-4.

Tanaka, T., 2000. The state transition model of the substorm onset. J. Geophys. Res. 105 (A9), 21081–21096. doi:10.1029/2000JA900061.

Tanaka, T., 2015. Substorm auroral dynamics reproduced by advanced global magnetosphere-ionosphere (M-I) coupling simulationAuroral Dynamics and Space Weather. Wiley, Hoboken, pp. 177–190.

Tanaka, Y.-M., Nishiyama, T., Kadokura, A., Ozaki, M., Miyoshi, Y., Shiokawa, K., Oyama, S.-I., Kataoka, R., Tsutsumi, M., Nishimura, K., Sato, K., Kasahara, Y., Kumamoto, A., Tsuchiya, F., Fukizawa, M., Hikishima, M., Matsuda, S., Matsuoka, A., Shinohara, I., Nose, M., Nagatsuma, T., Shinohara, M., Fujimoto, A., Teramoto, M., Nomura, R., Yukimatu, A.S., Hosokawa, K., Shoji, M., Latteck, R., 2019. Direct comparison between magnetospheric plasma waves and polar mesosphere winter echoes in both hemispheres. J. Geophys. Res. 124, 9626–9639. doi:10.1029/2019JA026891.

Tapping, K.F., 2013. The 10.7 cm solar radio flux (F10.7). Space Weather 11, 394–406. doi:10.1002/swe/20064.

Terada, K., Yokota, S., Saito, Y., Kitamura, N., Asamura, K., Nishino, M.N., 2017. Biogenic oxygen from Earth transported to the Moon by a wind of magnetospheric ions. Nat. Astron. 1 Article number: 0026.

Terada, N., Kulikov, Y.N., Lammer, H., Lichtenegger, H.I.M., Tanaka, T., Shinagawa, H., Zhang, T., 2009. Atmosphere and water loss from early Mars under extreme solar wind and extreme ultraviolet conditions. Astrobiology 9 (1), 55–70.

Teramoto, M., Hori, T., Saito, S., Miyoshi, Y., Kurita, S., Higashio, N., Matsuoka, A., Kasahara, Y., Kasaba, Y., Takashima, T., Nomura, R., Nose, M., Fujimoto, A., Tanaka, Y., Shoji, M., Tsugawa, Y., Shinohara, M., Shinohara, I., Blake, J., Fennell, J., Claudepierre, S., Turner, D., Kletzing, C., Sormakov, D., Troshichev, O., 2019. Remote detection of drift resonance between energetic electrons and ULF waves: multi-satellite coordinated observation by Arase and Van Allen Probes. Geophys. Res. Lett. 46, 11642–11651. doi:10.1029/2019GL084379.

Terasawa, T., et al., 1998. Observation of a cosmic-ray-modified interplanetary shock, Proc. 26th Int. Cosmic Ray Conf. (Salt Lake City), 6, 1998.

Thayer, J.P., Tobiska, W.K., Pilinski, M.D., Sutton, E.K., 2021. Remaining issues in upper atmosphere satellite drag. In: Coster, A.J., Erickson, P.J., Lanzerotti, L.J., Zhang, Y., Paxton, L.J. (Eds.), Space Weather Effects and Applications. https://doi.org/10.1002/9781119815570.ch5.

Thorne, R.M., et al., 2013. Rapid local acceleration of relativistic radiation-belt electrons by magnetospheric chorus. Nature 504, 411–414.

Toriumi, S., Wang, H., 2019. Flare-productive active regions. Living Rev. Sol. Phys. **16**, 3. https://doi.org/10.1007/s41116-019-0019-7.

Torok, T., et al., 2018. Sun-to-Earth MHD Simulation of the 2000 July 14 "Bastille Day". Eruption. Ap. J. 856, 75. https://doi.org/10.3847/1538-4357/aab36d.

Tousey, R., 1973. The Solar Corona. In: Rycroft, MJ, Runcorn, SK (Eds.), Space Research XIII. AkademieVerlag, Berlin, pp. 713.

Townsend, 2020. Space weather on the Moon. Phys. Today 73 (3), 66. https://doi.org/10.1063/PT.3.4438.

Tsunakawa, H., Takahashi, F., Shimizu, H., Shibuya, H., Matsushima, M., 2015. Surface vector mapping of magnetic anomalies over the Moon using Kaguya and Lunar Prospector observations. J. Geophys. Res. Planets 120, 1160–1185. doi:10.1002/2014JE004785.

Tsubouchi, K., Omura, Y., 2007. Long-term occurrence probabilities of intense geomagnetic storm events. Space Weather 5, S12003. doi:10.1029/2007SW000329.

Tsurutani, B.T., Gonzalez, W.D., Lakhina, G.S., Alex, S., 2003. The extreme magnetic storm of 1-2 September, 1859. J. Geophys. Res. 108 (A7), 1268. doi:10.1029/2002JA009504.

Tsurutani, B.T., et al., 2006. Corotating solar wind streams and recurrent geomagnetic activity: a review. J. Geophys. Res. 111, A07S01. doi:10.1029/2005JA011273.

Tsurutani, B.T., Verkhoglyadova, O.P., Mannucci, A.J., Lakhina, G.S., Huba, J.D., 2012. Extreme changes in the dayside ionosphere during a Carrington-type magnetic storm. J. Space Weather Space Clim. 2, A05. doi:10.1051/swsc/2012004.

Tsurutani, B.T., Lakhina, G.S., 2013. An extreme coronal mass ejection and consequences for the magnetosphere and Earth. Geophys. Res., Lett. 41, 287–292. doi:10.1002/2013GL058825.

Tsurutani, B.T., Hajra, R., Echer, E., Gjerloev, J.W., 2015. Extremely intense (SML < -2500 nT) substorms: isolated events that are externally triggered? AnGeo Comm 33, 519–524.

Tsyganenko, N.A., 1989. A magnetospheric magnetic field model with a warped tail current sheet. Planet. Space Sci. 37 (1), 5–20.

Tsyganenko, N.A., Stern, D.P., 1996. Modeling the global magnetic field of the large-scale Birkeland current systems. J. Geophys. Res. 101 (A12), 27187–27198. doi:10.1029/96JA02735.

Tsyganenko, N.A., Sitnov, M.I., 2005. Modeling the dynamics of the inner magnetosphere during strong geomagnetic storms. J. Geophys. Res. 110, A03208. doi:10.1029/2004 JA010798.

Tu, L., Johnstone, Colin P., C., P., Gudel, M., Lammer, H., 2015. The extreme ultraviolet and X-ray Sun in Time: High-energy evolutionary tracks of a solar-like star. A&A 577, L3. https://doi.org/10.1051/0004-6361/201526146.

Tylka, A.J., Adams Jr., J.H., Boberg, P.R., Brownstein, B., Dietrich, W.F., Flueckiger, E.O., Petersen, E.L., Shea, M.A., Smart, D.F., Smith, E.C., 1997. CREME96: a revision of the cosmic ray effects on micro-electronics code. IEEE Trans. Nucl. Sci. 44 (6), 2150–2160.

Tylka, A.J., et al., 2005. Shock geometry, seed populations, and the origin of variable elemental composition at high energies in large gradual solar particle events. Astrophys. J. 625, 474.

Ueno, H., Nakahira, S., Kataoka, R., et al., 2019. Radiation dose during relativistic electron precipitation events at the International Space Station. Space Weather 17. https://doi.org/10.1029/2019SW002280.

Usoskin, I.G., Alanko-Huotari, K., Kovaltsov, G.A., Mursula, K., 2005. Heliospheric modulation of cosmic rays: monthly reconstruction for 1951-2004. J. Geophys. Res. 110, A12108. doi:10.1029/2005JA011250.

Usoskin, I.G., et al., 2015. The Maunder minimum (1645-1715) was indeed a grand minimum: a reassessment of multiple datasets. A&A 581 (A95), 19.

Usoskin, I.G., Koldobskiy, S.A., Kovaltsov, G.A., Rozanov, E.V., Sukhodolov, T.V., Mishev, A.L., Mironova, I.A., 2020. Revisited reference solar proton event of 23 February 1956: assessment of the cosmogenic-isotope method sensitivity to extreme solar events. J. Geophys. Res. Space Phys. 125, e2020JA027921. https://doi.org/10.1029/2020JA027921.

Usoskin, I.G., Kromer, B., Ludlow, F., Beer, J., Friedrich, M., Kovaltsov, G.A., Solanki, S.K., Wacker, L., 2013. The AD775 cosmic event revisited: the Sun is to blame. A&A 552, L3. https://doi.org/10.1051/0004-6361/201321080.

Vainio, R., 2003. On the generation of Alfven waves by solar energetic particles. Astron. Astrophys. 406, 735.

Vainio, R., Desorgher, L., Heynderickx, D., Storini, M., Fluckiger, E., Horne, R.B., Kovaltsov, G.A., Kudela, K., Laurenza, M., McKenna-Lawlor, S., Rothkaehl, H., Usoskin, I.G., 2009. Dynamics of the Earth's particle radiation environment. Space Sci. Rev. 147, 187.

Van Allen, J., Frank, L., 1959. Radiation around the Earth to a radial distance of 107,400 km. Nature 183, 430–434. https://doi.org/10.1038/183430a0.

Van Allen, J.A., Fennell, J.F., Ness, N.F., 1971. Asymmetric access of energetic solar protons to the Earth's north and south polar caps. J. Geophys. Res. 76 (19), 4262–4275.

Vaisberg, O.L., Zastenker, G.N., 1976. Solar wind and magnetosheath observations at Earth during August 1972. Space Sci. Rev., 687–702.

Vasyliunas, V.M., 2011. The largest imaginable magnetic storm. J. Atmos. Sol. Terr. Phys. 73, 1444.

Vasyliunas, V.M., 2013. Time scale of the largest imaginable magnetic storm. Nonlin. Processes Geophys. 20, 19–23.

Vernov, S.N., Chudakov, A.E., 1960, Investigations of Cosmic Radiation and of the Terrestrial Corpuscular Radiation by Means of Rockets and Satellites. Sov. Phys. Usp. 3, 230.

Verronen, P.T., Santee, M.L., Manney, G.L., Lehmann, R., Salmi, S.-M., Seppala, A., 2011. Nitric acid enhancements in the mesosphere during the January 2005 and December 2006 solar proton events. J. Geophys. Res. 116, D17301. doi:10.1029/2011JD016075.

Vourlidas, A., Howard, R.A., Esfandiari, E., Patsourakos, S., Yashiro, S., Michalek, G., 2010. Comprehensive analysis of coronal mass ejection mass and energy properties over a full solar cycle. Astrophys. J. 722, 1522–1538. https://doi.org/10.1088/0004-637X/722/2/1522.

Walt, M., 1994. Introduction to Geomagnetically Trapped Radiation. Cambridge University Press, Cambridge, England.

Wang, Y.-M., Sheeley Jr., N.R., 1990. Solar wind speed and coronal flux-tube expansion. Astrophys. J. 355, 726–732.

Wang, Y., Zhang, J., 2007. A comparative study between eruptive X-class flares associated with coronal mass ejections and confined X-class flares. Astrophys. J. 665, 1428–1438. https://doi.org/10.1086/519765.

Watari, S., Kunitake, M., Watanabe, T., 2001. The Bastille Day (14 July 2000) event in historical large Sun-Earth connection events. Sol. Phys. 204, 425–438. https://doi.org/10.1023/A:1014273227639.

Watari, S., Nakamura, S., Ebihara, Y., 2021. Measurement of geomagnetically induced current (GIC) around Tokyo, Japan. Earth Planets Space 73, 102. https://doi.org/10.1186/s40623-021-01422-3.

Weatherwax, A.T., Rosenberg T.J., Lanzerotti L.J., Maclennan C.G., Frey H.U., Mende S.B., 2000. The distention of the magnetosphere on May 11, 1999: high latitude Antarctic observations and comparisons with low latitude magnetic and geopotential data, Geophys. Res. Lett. 27 (24), 4029-4032, https://doi.org/10.1029/2000GL000090

Weimer, D.R., 1995. Models of high-latitude electric potentials derived with a least error fit of spherical harmonic coefficients. J. Geophys. Res. 100 (A10), 19595–19607. doi: 10.1029/95JA01755.

Weimer, D.R., 2005. Improved ionospheric electrodynamic models and application to calculating Joule heating rates. J. Geophys. Res. 110, A05306. doi:10.1029/2004JA010884.

Weiss, B.P., Fong, L.E., Vali, H., Lima, E.A., Baudenbacher, F.J., 2008. Paleointensity of the ancient Martian magnetic field. Geophys. Res. Lett. 35, L23207. https://doi.org/10.1029/2008GL035585.

Welling, D.T., Love, J.J., Rigler, E.J., Oliveira, D.M., Komar, C.M., Morley, S.K., 2021. Numerical simulations of the geospace response to the arrival of an idealized perfect interplanetary coronal mass ejection. Space Weather 19, e2020SW002489. https://doi.org/10.1029/2020SW002489.

Wicht, J., Tilgner, A., 2010. Theory and modeling of planetary dynamos. Space Sci. Rev. 152, 501–542. doi:10.1007/s11214-010-9638-y.

Wiegelmann, T., Sakurai, T., 2012. Solar force-free magnetic field. Living Rev. Solar Phys. 9, 5.

Wijsen, N., Aran, A., Pomoell, J., Poedts, S., 2019. Modelling three-dimensional transport of solar energetic protons in a corotating interaction region generated with EUHFORIA. Astron. Astrophys. 622, A28.

Wold, A.M., Mays, M.L., Taktakishvili, A., Jian, L.K., Odstrcil, D., MacNeice, P., 2018. Verification of real-time WSA-ENLIL+Cone simulations of CME arrival-time at the CCMC from 2010 to 2016. J. Space Weather Space Clim. 8, A17. https://doi.org/10.1051/swsc/2018005.

Wood, B.E., Muller, H.-R., Zank, G.P., Linsky, J.L., 2002. Measured mass-loss rates of solar-like stars as a function of age and activity. Astrophys. J. 574, 412–425.

Wood, B.E., Muller, H.-R., Zank, G.P., Linsky, J.L., Redfield, S., 2005. New mass-loss measurements from astrospheric Lya absorption. Astrophys. J. Lett. 628, L143–L146.

Wordsworth, R., Forget, F., Millour, E., Head, J.W., Madeleine, J.-B., Charnay, B., 2013. Global modelling of the early martian climate under a denser CO_2 atmosphere: water cycle and ice evolution. Icarus 222, 1–19.

Wrenn, G.L., 1995. Conclusive evidence of internal dielectric charging anomalies on geosynchronous communications spacecraft. J. Spacecraft Rockets 32 (3), 514–520.

Yashiro, S., Gopalswamy, N., Michalek, G., St. Cyr, O.C., Plunkett, S.P., Rich, N.B., Howard, R.A., 2004. A catalog of white light coronal mass ejections observed by the SOHO spacecraft. J. Geophys. Res. 109, A07105. doi:10.1029/2003JA010282.

Yokoyama, N., Kamide, Y., Miyaoka, H., 1998. The size of the auroral belt during magnetic storms. Ann. Geophysicae 16, 566–573.

Yokoyama, T., Shibata, K., 1998. A two-dimensional magnetohydrodynamic simulation of chromospheric evaporation in a solar flare based on a magnetic reconnection model. Astrophys. J. 494, L113.

Zank, G., 1999. Interaction of the solar wind with the local interstellar medium: a theoretical perspective. Space Sci. Rev. 89, 413–688. https://doi.org/10.1023/A:1005155601277.

Zhang, S., Wimmer-Schweingruber, R.F., Yu, J., Wang, C., Fu, Q., Zou, Y., Sun, Y., Wang, C., Hou, D., Bottcher, S.I., Burmeister, S., Seimetz, L., Schuster, B., Knierim, V., Shen, G., Yuan, B., Lohf, H., Guo, J., Xu, Z., Freiherr von Forstner, J.L., Kulkarni, S.R., Xu, H., Xue, C., Li, J., Zhang, Z., Zhang, H., Berger, T., Matthia, D., Hellweg, C.E., Hou, X., Cao, J., Chang, Z., Zhang, B., Chen, Y., Geng, H., Quan, Z., 2020. First measurements of the radiation dose on the lunar surface. Sci. Adv. 6, eaaz1334.

Zhang, J., Temmer, M., Gopalswamy, N., et al., 2021. Earth-affecting solar transients: a review of progresses in solar cycle 24. Prog. Earth Planet Sci. 8, 56. https://doi.org/10.1186/s40645-021-00426-7.

Ziegler, J.F., 1999. The Stopping of Energetic Light Ions in Elemental Matter. J. Appl. Phys /Rev. Appl. Phys. 85, 1249–1272.

Further reading

Angenheister, G., Westland, C.J., 1921a. The magnetic storm of May 13-14, 1921: observations at Samoa Observatory. New Zealand, J. Sci. Technol. 4, 201–202.

Angenheister, G., Westland, C.J., 1921b. The magnetic storm of May 13-16, 1921, at Apia Observatory, Samoa. J. Geophys. Res. 26 (1, 2), 30–31. https://doi.org/10.1029/TE026i001p00030.

Bornebusch, J.P., Wissing, J.M., Kallnrode, M.-B., 2010. Solar particle precipitation into the polar atmosphere and their dependence on hemisphere and local time. Adv. Space Res. 45, 632–637.

Burton, R.K., McPherron, R.L., Russell, C.T., 1975. An empirical relationship between interplanetary conditions and Dst. J. Geophys. Res. 80 (31), 4204–4214. doi:10.1029/JA080i031p04204.

Chapman, S., Ferraro, V.C.A., 1931. A new theory of magnetic storms. Terr. Magn. Atmos. Electr. 36 (3), 171–186. doi:10.1029/TE036i003p00171.

Cliver, E.W., Hayakawa, H., Love, JJ., Neidig, D.F., 2020a. On the size of the flare associated with the solar proton event in 774 AD. Astrophys. J. 903, 41.

Cliver, E.W.F.M, Muscheler, R., 2020b. Solar longitude distribution of high-energy proton flares: fluences and spectra. Astrophys. J. Lett. 900 (1), L11 Number.

Fazio, G., Jelly, J., Charman, W., 1970. Generation of Cherenkov light flashes by cosmic radiation within the eyes of the Apollo astronauts. Nature 228, 260–264. https://doi.org/10.1038/228260a0.

Freiherr von Forstner, J.L., Guo, J., Wimmer-Schweingruber, R.F., Temmer, M., Dumbovic, M., Veronig, A., et al., 2019. Tracking and validating ICMEs propagating toward Mars using STEREO Heliospheric Imagers combined with Forbush decreases detected by MSL/RAD. Space Weather 17, 586–598. https://doi.org/10.1029/2018SW002138.

Futaana, Y., Barabash, S., Wieser, M., Lue, C., Wurz, P., Vorburger, A., Bhardwaj, A., Asamura, K., 2013. Remote energetic neutral atom imaging of electric potential over a lunar magnetic anomaly. Geophys. Res. Lett. 40, 262–266. doi:10.1002/grl.50135.

Gopalswamy, N., Yashiro, S., Krucker, S., Stenborg, G., Howard, R.A., 2004. Intensity variation of large solar energetic particle events associated with coronal mass ejections. J. Geophys. Res. 109, A12105. doi:10.1029/2004JA010602.

Hellweg, C.E., Baumstark-Khan, C., 2007. Getting ready for the manned mission to Mars: the astronauts' risk from space radiation. Naturwissenschaften 94, 517–526. doi:10.1007/s00114-006-0204-0.

Hodgson, R., 1960. On a curious appearance seen in the Sun. Mon. Not. R. Astron. Soc. 20, 15–17.

van der Holst, B., Sokolov, I.V., Meng, X., Jin, M., Manchester, W.B., IV, G.T., Gombosi, T.I., 2014. Alfven wave solar model (AWSoM): coronal heating. Astrophys. J. 782 (2), 81 Number.

Hoque, S.N.M.A., Alvi, A., Mahmud, Md.R., Sakhawat, S., Bhuiyan, M.I., 2021. On the use of directional derivatives of SuperDARN convection flow for estimating open-closed field line boundary during the intervals of substorm activity on 14 October 2001. Pol. Sci. 28, 100655. https://doi.org/10.1016/j.polar.2021.100655.

Kataoka, R., Miyoshi, Y., 2010. Why are relativistic electrons persistently quiet at geosynchronous orbit in 2009? Space Weather 8, S08002. doi:10.1029/2010SW000571.

Kataoka, R., Miyahara, H., Steinhilber, F., 2012a. Anomalous 10Be spikes during the Maunder Minimum: possible evidence for extreme space weather in the heliosphere. Space Weather 10, S11001. doi:10.1029/2012SW000835.

Kataoka, R., Nishiyama, T., Tanaka, Y.-M., Kadokura, A., Uchida, H.A., Ebihara, Y., Ejiri, M.K., Tomikawa, Y., Tsutsumi, M., Sato, K., Miyoshi, Y., Shiokawa, K., Kurita, S., Kasahara, Y., Ozaki, M., Hosokawa, K., Matsuda, S., Shinohara, I., Takashima, T., Sato, T., 2019. Transient ionization of the mesosphere during auroral breakup: Arase satellite and ground-based conjugate observations at Syowa Station. Earth Planet Space 71, 9. https://doi.org/10.1186/s40623-019-0989-7.

Kataoka, R., et al., 2020a. Plasma waves causing relativistic electron precipitation events at International Space Station: Lessons from conjunction observations with Arase satellite. J. Geophys. Res. Space Phys. 125, e2020JA027875. https://doi.org/10.1029/2020JA027875.

Kim, K.-C., Lee, D.-Y., 2014. Magnetopause structure favorable for radiation belt electron loss. J. Geophys. Res. Space Physics 119, 5495–5508. doi:10.1002/2014JA019880.

Kivelson, M.G., Ridley, A.J., 2008. Saturation of the polar cap potential: inference from Alfven wing arguments. J. Geophys. Res. 113, A05214. doi:10.1029//2007JA012302.

Kubo, Y., Kataoka, R., Sato, T., 2015. Interplanetary particle transport simulation for warning system for aviation exposure to solar energetic particles. Earth Planets Space 67, 117. doi:10.1086/s40623-015-0260-9.

Kunitomi, S.I., 1921. The magnetic storm of May 13-17, 1921 as observed at the Kakioka Magnetic Observatory. J. Meteorol. Soc. Japan. Ser. I 40 (7), en3–en6. https://doi.org/10.2151/jmsj1882.40.7_en3.

Love, J.J., Hayakawa, H., Cliver, E.W., 2019a. On the intensity of the magnetic superstorm of September 1909. Space Weather 17, 37–45. https://doi.org/10.1029/2018SW002079.

Love, J.J., Hayakawa, H., Cliver, E.W., 2019b. Intensity and impact of the New York Railroad superstorm of May 1921. Space Weather 17. https://doi.org/10.1029/2019SW002250.

Märki, A., 2020. Radiation analysis for Moon and Mars Missions. Int. J. Astrophys. Space Sci. 8 (3), 16–26.

Nagatsuma, T., Kataoka, R., Kunitake, M., 2015. Estimating the solar wind conditions during an extreme geomagnetic storm: a case study of the event that occurred on March 13-14, 1989. Earth Planet Space 67, 78. doi:10.1186/s40623-015-0249-4.

Owens, M.J., Lockwood, M., Barnard, L.A., et al., 2021. Extreme space-weather events and the solar cycle. Sol. Phys. 296, 82. https://doi.org/10.1007/s11207-021-01831-3.

Palmerio, E., et al., 2021. Investigating remote-sensing techniques to reveal stealth coronal mass ejections. Front. Astron. Space Sci. 8, 695966. https://doi.org/10.3389/fspas.2021.695966.

Saito, Y., Nishino, M.N., Fujimoto, M., et al., 2012. Simultaneous observation of the electron acceleration and ion deceleration over lunar magnetic anomalies. Earth Planet Sp 64, 4. https://doi.org/10.5047/eps.2011.07.011.

Saito, S., Kurita S., Miyoshi Y., Kasahara S., Yokota S., Keika K., Hori T., Kasahara Y., Matsuda S., Shoji M., Nakamura S., Matsuoka A., Imajo S., Shinohara I., 2021. Data-driven

simulation of rapid flux enhancement of energetic electrons with an upper-band whistler burst, 126, e2020JA028979, doi:10.1029/2020JA028979.
Sedov, L.I., 1959. Similarity and Dimensional Methods in Mechanics. Inforserach Ltd. Academic Press, Cambridge, Massachusetts, pp. 146.
Sukhodolov, T., Usoskin, I., et al., 2017. Atmospheric impacts of the strongest known solar particle storm of 775 AD. Sci. Rep. 7, 45257. doi:10.1038/srep45257.
Suzuki, T.K., Imada, S., Kataoka, R., Kato, Y., Matsumoto, T., Miyahara, H., Tsuneta, S., 2013. Saturation of stellar winds from young Suns. Publ. Astron. Soc. Japan 65, 98. doi:10.1093/pasj/65.5.98.
Tanaka, T., 2000. The state transition model of the substorm onset. J. Geophys. Res. 105 (A9), 21081–21096. doi:10.1029/2000JA900061.
Uchida, T., 1909. Geomagnetically induced currents on September 25, 1909. J. Inst. Electr. Eng. Japan 29 (255), 701–721.
Usoskin, I.G., Kovaltsov, G.A., 2021. Mind the gap: new precise 14C data indicate the nature of extreme solar particle events. Geophys. Res. Lett. 48, e2021GL094848. https://doi.org/10.1029/2021GL094848.
Viljanen, A., Myllys, M., Nevanlinna, H., 2014. Russian geomagnetic recordings in 1850-1862 compared to modern observations. J. Space Weather Space Clim. 4, A11. doi:10.1051/swsc/2014008.
Willis, D.M., Stephenson, F.R., Fang, H., 2007. Sporadic aurorae observed in East Asia. Ann. Geophys. 25, 417–436.
Winckler, J.R., Peterson, L., Hoffman, R., Arnoldy, R., 1959. Auroral X-rays, cosmic rays, and related phenomena during the storm of February 10-11, 1958. J. Geophys. Res. 64 (6), 597–610.
Zeitlin, C., Schwadron, N.A., Spence, H.E., Jordan, A.P., Looper, M.D., Wilson, J., et al., 2019. Update on galactic cosmic ray integral flux measurements in lunar orbit with CRaTER. Space Weather 17, 1011–1017. https://doi.org/10.1029/2019SW002223.

Index

Page numbers followed by "*f*" and "*t*" indicate, figures and tables respectively.

A

Air shower, 27
  energy deposition, 40
  nuclear spallation reactions, 27
  schematic diagram, 28*f*
Akasofu's epsilon, 100
Akatsuki spacecraft, 136
Alfvén Mach number, 50, 51
Alfvén waves
  electromagnetic and hydrodynamic waves, 7
  solar wind, 7*f*
  wing formation, 51
Ampere's law, 25
Anomalous cosmic rays (ACR), 119
Appleton-Hartlee equation, 66
Atmosphere, 19
Atmospheric
  chemical reactions, 40
  ionization, 28*f*, 78
Aurora Borealis, 55*f*
Auroral electrojet (AE) index, 48
Auroral oval, 1
Auroral substorm, 45
  ionosphere, 45–46
  schematic diagram, 45*f*
Auroral zone, 14*f*
Auroras
  colors, 20–22
  diffuse, 19, 21*f*
  discrete, 19 *See also* (Discrete aurora)
  DMSP satellite, 23*f*
  drawing of, 20*f*
  fan-shaped, 54*f*
  spectra, 23*f*

B

Bohm limit, 40
Boltzmann
  constant, 26–27
  equation, 105
Bounce-averaged gradient-curvature drift, 60

C

Calorimetric Electron Telescope (CALET), 10
CALS model, 13
Cancer death, probability, 129
Carrington-class magnetic storm, 55, 82, 88
Carrington flare amplitude, 32–34
Carrington magnetic storm, 52*f*
Carrington storm, 52, 71
Chapman-Ferraro current, 16
Charged particles, phase-space distribution function, 116–118
Charge-exchange collisions, 102–103
Chromosphere, 1–3
Cold plasma, 18
Combined Release Radiation Effects Satellite (CRRES) satellite data, 62, 63
Comprehensive Inner- Magnetosphere Ionosphere (CIMI) model, 105
Comprehensive Ring Current Model (CRCM), 105
Continuous pulsations (Pc), 47
Coronal holes, 4
Coronal loops, 93
Coronal mass ejections (CMEs), 31, 35
  amplitude, 72
  arrival, 96
  corotating interaction regions (CIRs), 59, 60*f*
  cumulative distribution function, 86*f*
  earth-directed, 35
  Halloween storms, modeled propagation, 97*f*
  interplanetary shock, 35
  magnetic clouds, 35
  magnetic field, 38
  magnetic storms, 49–50
  outer belt electron flux, depletions, 57–59
  red-sign events, 56
  schematic diagrams, 36*f*

Coronal mass ejections (CMEs) (*Continued*)
solar counterpart, 35
solar energetic particle, 39*f*
solar wind density, 38
solar wind parameters, 38
Corotating interaction region (CIR), 4, 6*f*, 49–50
Cosmic rays, 27 *See also* Galactic cosmic rays
flux, simulated curves/observed data points, 119*f*
modified shocks, 38
nuclei, local interstellar spectrum, 115–116
Cosmic Ray Telescope for the Effects of Radiation (CRaTER), 130–131
Cosmogenic radionuclides, 42–43
Coulomb collisions, 19
Cross-energy coupling process, 108–109
Cross-energy/cross-region couplings, sketch, 106*f*
Cross polar cap potential (CPCP), 51
Crustal magnetic anomalies, 126
Cumulative distribution function, 85–86
Current-carrying electrons, 74–75

D

Dark filaments, 35
Daytime X-rays, 22–24
Deep dielectric charging, 78–79
Deep Flare Net, 93
Dellinger effect, 65, 78
Dessler-Parker-Sckopke equation, 53–54
Diamagnetic current, 16
Diffusion coefficient, 110
Diffusive shock-acceleration, 10–12
Disappearing filament, 35
Discrete aurora
energy spectrum, 19–20
model spectra, 21*f*
Drift
echoes, 62–63
kinetic equation, 105
resonance, 108–109
velocity, 25
Dst correction, 16
Dst index, 48, 53–54, 81

E

Earth
dipole-like magnetosphere, 123
magnetosphere, 17*f*
radius, 13
Effective dose rate, global map, 111*f*
Electric conductivity, 6–7
Electromagnetic bursts, 32*f*, 66*f*
Electromagnetic wave transmission, 25
Electron density, altitude profiles, 23*f*
Electron-neutron collision frequency, 65
Electrons
Larmor radius, 15*f*
steady-state momentum equations, 24–25
Electrostatic charging, 131
Empirical space weather forecast, schematic picture, 108*f*
Energetic storm particles, 122*f*
and reservoir, 122*f*
solar energetic particle flux, 120–122
Energy conservation equation, 98
ENLIL model, 96, 98, 124–125
Equatorial electrojet, 70
Equivalent current, 25
Ground-based HF radars, 25
Hall current, 25
Eruptive flux tube, simulated evolution, 94*f*
ESA Mars Express mission, 91
European Heliospheric Forecasting Information Asset (EUHFORIA), 98
Extravehicular activities (EVA), 77
Extreme UV (EUV), 35

F

Faint-Young-Sun paradox, 132
Faraday's law, 71, 98
Field-aligned current (FAC), 16
Chapman-Ferraro current, 16
ionospheric pattern, 18*f*
Filaments
eruption, 35
magnetic polarity/helicity of, 35
First adiabatic invariant, 59
Flickering auroras, 45–46

Focused transport equation, 112
Fokker-Plank equation, 107–108
Forbush decrease, 40, 120
Force-free magnetic field, 93, 94*f*
Forecasting
  evaluating metrics, 113
  ground level enhancement, 111
  magnetic storms, 102
  radiation belts, 106
  solar flares, 93
  substorms, 99
Fukushima's theory, 25

G

Galactic cosmic rays (GCR), 10, 115
  differential flux, 11*f*
  diffusion tensor, 116–118
  energy spectra, 10, 117*f*
  modeling efforts, 113
  predicting, 115
  solar cycle variation, 130–131
  solar modulation, 115–116
  troposphere, 27
Gauss coefficients, 13
Gaussian energy spectrum, 19–20
Gaussian width, 120
Geocentric Solar Ecliptic (GSE), 47–48
Geocentric Solar Magnetospheric System (GSM), 47–48
Geomagnetically induced currents (GIC), 71*f*, 104
Geomagnetic disturbance (GMD), 46
  effects, 67, 76*t*
  outstanding magnetic storms, 50*t*
  power outages, 70
  rate of change, 72*f*
  substorm-related auroral electrojet, 71–72
Geomagnetic field poses, 16–17
Geomagnetic-induced currents (GIC), 70
Geomagnetic pulsations, 47
Geomagnetic reversal, 13–14
Geostationary Operational Environment Satellite (GOES), 93
  solar flares, light curves, 33*f*
  spacecraft, 42
Geosynchronous magnetopause crossing (GMC), 16–17
Geosynchronous orbit (GEO), 16–17
  electron flux, 106
  surface charging, 74–75
Gerasimovic anomaly, 126
Giant arcades, 35
GIC Working Group of NASA, 73–74
Global muon detector network (GMDN), 96–97
Global Navigation Satellite System (GNSS), 66, 86–87
  ionospheric irregularities, 75–76
  satellites, 66
Global navigationsatellite systems(GPS)
  ionosphere/radio propagation, 67*f*
  Wide Area Augmentation System, 91
Gradual storms, 49–50
Grand minimum, 4
Granules, 1–3
Ground-based neutron monitors, 39–40
Ground level enhancement (GLE), 39–40, 43*t*, 111
  forecasting, 111
  time-integrated proton flux, 42
  typical flight altitude, in polar region, 77
Guiding-center approximation, 105
Gyrating particles, Störmer's theory, 15–16
Gyro motion, schematic diagram, 61*f*
Gyro radius, 14

H

Hale cycle, 3
Hall conductivity, 24–25
Hall current, 25
Halloween event, 37*f*
Halloween storm, 69
Halo coronal mass ejections (CMEs), 37*f*
Heliosphere, schematic illustration, 9*f*
Heliospheric current sheet (HCS), 8–9
  global evolution, 10
  heliopause, 8–9
  solar cycle 22, 10*f*
  solar dipole component, 118*f*
  tilt angle, 118
Himawari-3, at geosynchronous orbit, 77–78

Historical auroral events, timeline, 87*f*
Hydrogen wall, 8–9
Hydro-Quebec blackout, 67, 69
Hydrostatic equilibrium, 26–27

I
IGRF model, 13
Inertial current, 16
Intense magnetic storms, 103
International Geophysical Year (IGY), 56–57
International Space Station (ISS), 10, 62*f*
  Calorimetric Electron Telescope (CALET), 10
  dose rate observation data, 62*f*
Interplanetary magnetic field (IMF), 4, 8–9, 46
Interplanetary scintillation (IPS), 96–97
Ionosphere, 19, 67*f*, 74
  ionospheric conductivity, 32–34
  ionospheric convection, 25, 26*f*
  ionospheric density, 24
Ion-pair production, 40
Irregular pulsations (Pi), 47

J
Japanese ASCA satellite, 74
Joule heating, 74

K
Killer electrons, 78–79
Kirchhoff's law, 24

L
Larmor radius, 15*f*
Local interstellar spectrum (LIS), 115–116
Loss-of-equilibrium, 94
Low-Earth orbit (LEO), 29
Lunar Reconnaissance Orbiter, 130–131

M
Mach number, 10–12, 50–51
Magnetic
  anomaly regions, 127*f*
  clouds, 35
  configuration, 94–95
  divergence-free nature of, 13
  drift, 118, 119
  drift velocity, 60
  half-day time profile, 69*f*
  map, 137*f*
  permeability, 71
  reconnection, 31
  reversal, 13–14
  rigidity, 14
Magnetic storms, 4, 35, 47, 48, 49–50, 67–69, 76, 90, 102, 103
  amplitude of, 53–54, 55, 82
  Carrington-class, 82
  cumulative distribution function, 83*f*
  Dst amplitude, 73*f*
  Dst index, 53*f*
  forecasting, 102
  main phase, 47
  observation data *vs.* simulation data, 107*f*
  recovery phase, 47
  solar cycle 23, 107*f*
Magnetohydrodynamic (MHD) simulations, 1, 6, 96
  convection patterns, 101*f*
  field-aligned current (FAC), 101*f*
  frozen-in condition, 8
  global MHD simulation, 100
  nonthermal components, modeling works, 124
  real-time global simulation, 100–101
  shocks on Earth, 102*f*
  solar wind simulations, 98
Magnetopause, 16
Magnetosheath, 17*f*
Magnetosphere, 16, 17*f*
Magnetotelluric (MT), 71
Marian atmospheric evolution, schematic illustration, 133*f*
Mars
  atmosphere energy deposition rates, 134*f*
  elliptical orbit, 131
  global diffuse aurora, 137*f*
  proton fluxes *vs.* time, 136*f*
  surface, 77
  thin atmosphere, 134
Mars Atmosphere and Volatile Evolution Mission (MAVEN), 131–132
Mars Global Reference Atmospheric Model 2001, 134
Mars Global Surveyor, 137*f*
Martian crust, magnetization, 18–19

Martian Radiation Environment Experiment (MARIE), 131–132
Maunder Minimum, 4
Maximum likelihood estimation, 81
Maxwell
  distribution, 19–20
  equations, 6–7, 105
  law, 6–7, 70, 98
Mesosphere, 26
Middle atmosphere, 25
Mirror point, 59
Monte-Carlo simulation, 126–128
Moon
  magnetic anomaly, 127*f*
  orbital radius, 126
  space weather, 125
  stations, 126

N
National Institute of Information and Communications Technology, 66
Negative ionospheric storms, 24
Neutral gas velocity, 24–25
Neutral nitrogen, emissions, 22
Neutron monitor network, solar modulation parameter, 117*f*
Neutron, spallation of, 43–44
New York Railroad systems, 72–73, 88
Nitrogen molecules, 22
NmF2, 24
NOAA/SWPC, 80–81
  extreme effects, 67*t*
  G-scale, 48
  R-scale, 33*t*
  solar flares, 66
  S-scale, 42*t*
Nonadiabatic processes, 59
Nonlinear force-free field, 93

O
Ohm's law, 8
Oulu cosmic ray station, 116*f*
Outer radiation belt electron flux, averaged variation, 60*f*
Ozone depletion, 135
Ozone destruction, 41–42
  high-latitude, 41*f*
  strong SEP events, 41–42

P
Parker equation, 116–118
Parker spirals, 8, 112
  schematic illustration, 6*f*
  solar wind, 47–48
Pedersen conductivity, 24–25
Phantom command, 79
PHITS simulation, 128*f*, 134
Photo-chemical reactions, 20–22
Photosphere, 1–3
Physic-based prediction scheme, 94–95
Plasma, 1
Plasma bubbles, 75–76
Plasma flow, 6–7
Plasma frequency, 65
  ionosphere, 65
  radio waves, 65
Plasmasphere, 18
Plasmoid, 45
Polar cap absorption (PCA), 78
Polarity-flipping time, 13–14
Positive ionospheric storms, 24
Potential field, 93
Power-law distribution, 81
Prediction
  galactic cosmic rays, 115
  prediction efficiency, 113
  solar energetic particles, 120
Probability of detection (POD), 113
Probability of false detection (POFD), 114
Proton intensity, 120
Protons, 15*f*

R
Radiation Assessment Detector (RAD), 135
Radiation belts, 4, 56, 57
  activity, 63
  current-carrying electrons, 74–75
  electron flux, 79
  energy density/pressure, 59
  forecasting, 106
  inner radiation belt protons, integral flux/energy spectra of, 58*f*
  outer belt electrons, enhancement of, 57–59
  particles, 110
  schematic illustration, 80*f*

Radio blackout impacts, 33*t*
Radio communication, 131–132
Radio propagation, 67*f*
Radio waves, refractive index, 66
Rankine-Hugoniot relationship, 17–18, 38
Real-time data, 93
Receiver operating characteristic (ROC), 114
Reconnection-triggered double-arc instability, schematic diagram, 95*f*
Red prominences, 35
Relativistic electron precipitations (REP), 110–111
Reservoir/invariant spectral region, 123
Resonant energy, 110
Resonant wave-particle, 110
Ring current, 47
  particle energy, 53–54
  schematic diagram of, 104*f*
Root-mean-square error (RMSE), cross-correlation efficient, 113
Russell-McPherron effect, 47–48

S

Schwabe cycle, 3
Sedov-Tayler solution, 12
Single event effect (SEE), 77
Single event upset (SEU), 77
Slot region, 62–63
Snowstorm effect, 78*f*
Solar activity, atmospheric temperature, 27*f*
Solar cycles
  electron flux at LANL-GEO/solar wind speed, 58*f*
  thermospheric temperature, 29
Solar Dynamic Observatory, 93
Solar eclipse, 35
Solar energetic particle (SEP), 39, 42, 42*t*, 123*f*
  asymmetry of, 41–42
  atmosphere, radiation shelter, 76–77
  coronal mass ejections (CMEs), 39*f*
  cumulative distribution function, 85*f*
  energy spectra, 44*f*
  high-flux SEP, 39–40
  intensity, streaming limit of, 121*f*
  lunar regolith, 129*f*
  ozone destruction, 41–42
  possible effects, 79*t*
  predicting, 120
  radiation dose rate, 135*f*
  snow storm, 77
  X-ray intensity, 44–45
Solar flares, 31, 84
  cumulative distribution function, 85*f*
  effect, 32–34
  electromagnetic wave bursts, 65
  light curves of, 33*f*
  magnitude, 31–32
  schematics diagram, 32*f*
  X-ray intensity, 31
Solar luminosity, 1–3
Solar magnetic field, 3
Solar modulation theory, 115, 119
Solar protons, 78–79
  event, 39
  pitch-angle distribution, 112
Solar radio bursts, 34, 65
  electromagnetic waves, 34
  schematic diagram, 34*f*
Solar system, 126*t*
  earth/giant planets, 19*t*
  satellites, 126*t*
  terrestrial environment, schematic illustration, 2*f*
Solar wind, 1, 4, 6
  Alfvén waves, 7*f*
  constant, 8
  dynamic pressure, 8, 9–10, 12
  energy flow, 46
  interaction, magnetic anomaly, 127*f*
  magnetosphere, 18, 100
  simulations, 97
  southward-directing magnetic field (SBZ), 49*f*
  speed, latitudinal variation, 5*f*
  sub-Alfvenic, 51
  supersonic plasma, 17–18
South Atlantic Anomaly (SAA), 60–62, 130–131
Southward-directing magnetic field (SBZ), 35, 49*f*, 99
  cloud, 48
  high-speed solar wind, 59
  shock downstream, 48

Space weather, 1
Bastille Day event, 90
Carrington event, 87
CME speed, 91
events, outstanding, 87
GLE event, 88
Halloween event, 90
high-speed/worst SEP event, 89
Hydro-Quebec blackout event, 89
Mars event, 92
MIR event, 89
New York Railroad event, 88
"perfect" storm, 92
STEREO event, 92
super SC event, 90
Tokyo aurora event, 88
vulnerable infrastructures phenomena, 68*f*
Space Weather Follow-On program, 96
Spectral hardness, 88
S-scale, 42
Stable auroral red (SAR) arc, 19, 104
Starlink satellite prototypes, 29
Stern models, 99
Stochastic differential equation (SDE) method, 112
Storm-enhanced density (SED), 76
Störmer's theory, 15–16
Storm sudden commencement (SSC), 47
Stratosphere, 26
Streaming limit, 120–122
Stream interfaces, 4, 106–107
Subauroral polarization stream (SAPS), 104
Substorms, 35, 45, 47–48
commencements, 16
energy dissipation, 46
forecasting, 99
magnetic storms, 47–48
NOAA G-scale, 48*t*
Sudden commencements (SCs), 16, 84*t*
amplitude of, 83*f*
substorms, 16
Sudden impulse, 47
Sun
flips, magnetic moment polarity of, 3
inner structure/atmosphere, 2*f*
primary ionize, 22–24
rotation period, 3
Sunspots, 1
butterfly diagram, 10*f*
defined, 1
explosive energy, flow diagram, 32*f*
schematic diagrams, 36*f*
sketch, 33*f*
space environment, 3
SuperDARN, 25, 26*f*
Super-fountain effect, 75–76
SuperMAG website, 83–84
cumulative distribution function, 84*f*
substorm list, 83–84
Supernovae, 10–12, 39
Supernova remnant, 12
Surface magnetic field, 93

T
Termination shock, 8–9
Terrestrial planets, characteristics, 131*t*
Thermosphere, 26
Total electron content (TEC), 24, 66
errors of, 66
upward-viewing GPS antenna, 75*f*
Troposphere, 26
Tsyganenko models, 99
Twin STEREO satellites, 96

U
Ultra-low frequency (ULF) waves, 47, 47*t*
Ulysses spacecraft, 4, 5*f*

V
Violent energy flow, atmospheric end-product of, 19

W
Wave-particle interactions, 111
Weimer models, 99
White light flares, 31, 32–34
Whole body irradiation, radiation effects in humans, 130*t*
Wide Area Augmentation System (WAAS), 91

FSC
www.fsc.org
MIX
Papier | Fördert gute Waldnutzung
FSC® C083411

Kolibri 360 GmbH
Ferdinand-Jühlke-Straße 7
99095 Erfurt, Deutschland
produktsicherheit@kolibri360.de